THE PEOPLE'S RIGHT TO KNOW:
Media, Democracy, and the Information Highway

The LEA Telecommunications Series
A Series of Volumes Edited by Christopher H. Sterling

Williams/Pavlik • The People's Right to Know: Media, Democracy, and the Information Highway

THE PEOPLE'S RIGHT TO KNOW: Media, Democracy, and the Information Highway

Frederick Williams
John V. Pavlik
Editors

The Freedom Forum
Media Studies Center
At Columbia University in the City of New York

1994

LAWRENCE ERLBAUM ASSOCIATES, PUBLISHERS
Hillsdale, New Jersey   Hove and London

Lawrence Erlbaum Associates, Publishers
365 Broadway
Hillsdale, New Jersey 07642

Library of Congress Cataloging-in-Publication Data

The People's right to know : media, democracy, and the information highway / Frederick Williams, John V. Pavlik, editors.
p. cm.
Includes bibliographical references and index.
ISBN 0-8058-1490-6. — ISBN 0-8058-1491-4 (pbk.)
1. Information society. 2. Democracy. 3. Freedom of information. 4. Mass media. I. Williams, Frederick, 1933- . II. Pavlik, John Vernon.
HM221.P427 1994
302.23—dc20 93-34308
CIP

Books published by Lawrence Erlbaum Associates are printed on acid-free paper, and their bindings are chosen for strength and durability.

Printed in the United States of America
10 9 8 7 6 5 4 3

CONTENTS

PART III:
POLICYMAKING REGARDING CITIZEN INFORMATION SERVICES

Cover art: Niculae Asciu, 1992

Preface

Many of the ideas expressed in this book would have seemed fanciful, if not fantastic, a decade ago. In a sense, what this book and the various activities—conferences and seminars—associated with it urges is a radical proposal for ultimate information literacy. On one side of the equation is the great storehouse of information assembled and accumulated by all of society's institutions and largely held by vendors in the communications industry and in nonprofit organizations. Stored information from the past as well as that being currently developed and some yet unanticipated is stockpiled in that imaginary warehouse. On the other side of the equation are institutions and organizations as well as citizens, consumers, and others who constitute the human community—all needing, wanting, and asking for information. In between these transactions are various systems, instruments, and technologies capable of delivering information, knowledge, and data to those who want and need it.

It is this ground between actual information and the demand for it that commands the attention of this book and its various contributors. On that ground it is proposed that a vast information highway be laid, a multifaceted system that can acquire, process, disseminate, and store information of all kinds. Students of technology tell us that the technical tools to make this happen are either already available or on the drawing board. These tools would allow individuals and institutions to plug into the information highway and retrieve any kind of information they might desire—information structured to meet their specific needs and responsive to interactive commands and messages. In one formulation, the information highway requires a vast fiber optic network that would augment and supplement the electronic and other messages now available through telecommunication aided by satellites and computers and through older technologies of printing.

What stands between realization of the dream that would grant to everyone the right to this vast network of electronic services? For one thing, money. In a capitalist system, money talks and defines who gets access and who does not. Those who can pay are already blessed with information wealth that includes a vast array of media and data sources coming into the home and office. Others get some of these benefits through such institutions as schools, libraries, and businesses, for example. However, most of the earth's peoples and even a strong majority in the developed information societies lack this opportunity. Another force blocking the information highway dream is government regulation and law. Institutional entities in the media are fighting a media war where the stakes are high for various media industries, such as newspapers, cable television, and the telephone companies. The so-called policy outcome of these disputes will determine the legal and regulatory framework for the information highway and will, of course, define the rights of individuals to information, currently uncharted territory.

Of course, the story of the information highway is more than a mechanistic examination of haves and have-nots, wherein freedom flourishes on one side of the scale and control on the other. It involves the media economy specifically and the general economy as well, both of which involve consumers. The needs and wants of people as expressed attitudinally to market researchers must be demonstrated more viscerally in actual buying decisions and in signals that a given new technology will have some takers, whether at the personal or corporate level or somewhere in between.

Very much at the forefront of the information highway debate is the matter of consequences. What, it is asked, will a true information highway—where most citizens enjoy a wide range of information services on demand—do to local communities, government, and business entities; other units of society; and democracy itself? And concurrently, how do we know? The idea of universal information literacy suggests that every human being in the society who seeks it can have basic information services delivered speedily and in a cost-effective fashion, if not free. Few considerations in society are more important than this one is to the future of human discourse, but rarely do society's leaders take clear stands on this complex and controversial concern. As a U. S. senator, Albert Gore introduced an information highways bill that would have charted some policy directions and provided a public service network. As vice president of the United States, he and President Bill Clinton have an opportunity to develop a plan that would lay the foundation for an information highway that mixes public and governmental concerns and those of the private sector. Clearly, government has a role to represent the public interest in the unfolding process, while industry and other players seek economic health for the system and the chance to make a profit or put forth their views and vested interests.

The information highways debate would be moot without the array of new technologies, hardware, and software, and a good many more on the drawing boards that remove most, if not all, of the earlier technical barriers. What is left to be decided is how the highway will proceed, where it will go, who will have access, what it will cost, what it will do to and for us, and a good deal more. The highway analogy is apt because the decisions being made now are not unlike those made about the telegraph and the railroad in the 19th century or about the telephone and telecommunications in the 20th century. The 21st century promises to be a time for people to factor themselves into the richest lode of information ever accumulated by human beings. Whether they are able to do it coherently on a level playing field or against mean odds that will further divide and fragment society is still not clear. Many of the instincts in the information highways concept are utopian and utopian ideas, if history serves, typically do not work. Still, in a post-Cold War world where there is potential for human cooperation without the presence of an East-West prism, and in the midst of a surge of democratization and liberalization of various paternalistic governments, the chance for a democratized information highway exists as it has never before.

This book is an attempt to define and sort out the information highway debate, to make it more understandable to specialists and ordinary citizens. The book grows out of a residential fellowship project at The Freedom Forum Media Studies Center at Columbia University begun in 1991 by University of Texas Professor Frederick Williams, then a senior fellow and a noted scholar in the field of communication and technology. In written work, roundtable discussions, and a national conference in November 1992, the information highway project has advanced at the Center. Here, associate director John Pavlik has guided the work and brought much of it to fruition. He and Professor Williams collaborated in editing this book as they have in other endeavors.

We at the Media Studies Center are proud to have been midwives to this new arrival on the communications scene and are privileged to track and trace its growth. This book is another step in that process, and it is our hope that the various views and perspectives found therein will provide a connecting thread and enrich the national and international conversation on the information highway.

— Everette E. Dennis
Executive Director
The Freedom Forum
Media Studies Center

ACKNOWLEDGMENTS

If advances in the nation's public telephone network are going to make information services as easy to use as ordinary voice calls, or newspapers argue that they have vast new electronic services awaiting their readers, why do we not devote more attention to the information needs and wants of everyday citizens? It is already well known that we have vast inequities across America's socioeconomic classes in access to information critical for everyday life in our increasingly multicultural and technology-driven society. Unfortunately, however, the public—especially the "information poor"—has been all but overlooked in national debates between telephone companies and newspaper publishers on who should provide services, not to mention their bias toward an upscale market looking for news headlines, stock market tips, and electronic shopping. If we, indeed, have a powerful new medium at our disposal, why not consider policy that in part attempts to close the information gap among our citizens—meaning job, education, and health information services; legal information on such topics as immigration; and transactional services that give assistance on such routine but time-consuming tasks as renewing a driver's license or registering to vote?

We have nearly achieved our goal of "universal service" in voice telephony; why not a 21st century upgrade to universal service in access to essential information and transactions? This has been the topic of ongoing research activities beginning with a policy paper developed during Frederick Williams's tenure as a senior fellow at the Freedom Forum Media Studies Center in the spring of 1991. Under the joint direction of John V. Pavlik, the Center's associate director for Research and Technology Studies, and Williams, the Center subsequently hosted a research roundtable in February 1992 and a national conference in October that are the main bases for this collection. The authors express their gratitude to Everette E. Dennis, executive director of The Freedom Forum

Media Studies Center, whose continuing support has made this book possible. The authors also thank Mark A. Thalhimer, the Center's technology manager, who has also contributed importantly to this project, both through his research on the background and history of the technology of the information highway and his efforts in the production of the book.

The authors also are indebted to Christopher H. Sterling, professor and director of the Telecommunications Program at George Washington University, editor of the Telecommunications Series launched by this book, and to Hollis Heimbouch, Lawrence Erlbaum Associates' senior editor responsible for the book.

Hopefully from *The People's Right to Know* we will have improved our basis for deciding if there are means by which an enhanced public telecommunications network can benefit the everyday, working American.

—Frederick Williams
—John V. Pavlik

Part I

Envisioning the Shape and Feel of a National Information Service

The first phase of the Center's research activities in electronic information services reflected mainly visions of examples or national policy to develop such services. Chapters 1 and 2 are examples of the "feel" of such services. Chapters 3 and 4 reflect a compilation of research issues and a roundtable examination of the national service concept. One major implication was that we needed to look much more at the users of such services, thus we turned more to a concept of "citizen" information service as reflected in the chapters in Part II.

Chapter 1

ON PROSPECTS FOR CITIZENS' INFORMATION SERVICES

BY FREDERICK WILLIAMS

MILAGROS SABE DE ESAS COSAS
["Milagros Knows About Those Things"]

"No sé si quiero hacer esto" [I don't know if I want to do this"] ... says a young Hispanic woman probably in her late teens; she is dressed in jeans and a large green imitation-leather coat; her hair is covered with a red bandana. She speaks into a handkerchief, *"... pero no sé que más puedo hacer"* ["but I don't know what else to do"] .

A second young woman in jeans, an army-like jacket and "Giants" baseball cap softly asks:

"¿Cómo lo encontraste?" ["How'd you find him?"]

"Milagros me dijo ... me dijo que necesitaría dinero en efectivo..." ["Milagros told me ... I'd need cash,"] *"... y que tomara el Express de la Avenida 7 a este lugar en el Bronx"* [... and to take the 7th Ave. Express to this place in the Bronx"] .

She holds out a slip of paper.

"Milagros sabe de esas cosas" [Milagros knows about those things"].

A third member of the party, dressed in a skirt and blouse mostly covered by a brown overcoat, gives directions: *"Vamos a la Calle 96 y allí cambiamos"* [We'll go down to 96th St. and transfer"]. Perhaps she is an older sister who is along to help.

This conversation is overheard on the platform deep beneath Broadway in the 168th street station of the New York City subway. Being so far underground is probably why this cavernous space is so poorly ventilated and overheated; the usual tinge of urine smell is always stronger here. It's about 9:30 in the morning when the commuter crowd has thinned out. But there is no place to sit because a cluster of homeless men and their bundles occupy the two wooden benches that the transit police will clear before noon. The young women keep their distance further up the platform. Their conversation gives way to the rumbling arrival of the downtown train.

"Si pudiera mantener mi bebe, me gustaría ..." ["If I could keep my baby, I'd wish ..."]

The scene stirs up memories from the previous day of coeds laughing and clowning while they build a snowman on the steps of Low Library at Columbia University. Unfortunately, the weighty decisions of life fall sooner on some in our society than others, perhaps most often on the least prepared. Our society should be able to do better.

* * *

INFORMATION RICH VERSUS INFORMATION POOR

The dilemma of these three anxious young women is much more than an everyday anecdote about getting an abortion in a society that still gives mixed signals on individual rights to social services. It is an increasingly frequent sign of the growing gap between the information rich and the information poor in our evermore complex, technology driven, so-called *information* society. Whether you first read of the increasing importance of information or knowledge as a new economic resource as described in a scholarly volume like Daniel Bell's *The Coming of Post-Industrial Society* or Alvin Toffler's more popularized, *The Third Wave,* it is abundantly clear that if you do not have access to this new resource, you will be doomed to remain in the underclass of what are otherwise visibly affluent societies.

Surely, education and access to "know-how" have always been a key to upward social mobility, but it seems that in our time with the structural changes in the U.S. post-industrial economy, the lack of these capabilities leaves no feasible alternatives. You cannot make a decent living with a strong back in an information society. In fact, like the friends of Milagros, the lack of critical information necessary to meet the problems of everyday life is the most dooming quality of being among the information underclass. Unfortunately, too, the growing underclass is dangerously correlated with multicultural divisions of our society.

Could it be that recent advances in the tools of the information society are not developing in forms to serve citizens who need them the most? Are advances in computer and telecommunications applications contributing to growth of the information gap in our society when with an enlightened national communications policy, they could be used to close that gap? Let us pursue answers to these important questions by outlining the case for a National Citizens' Information Service in this country.

ELECTRONIC INFORMATION SERVICES AS A NEW PUBLIC MEDIUM

Visualize for a moment, if you will, the availability of a new communications medium that could provide you with information, message or transactions services for most any need critical to your everyday survival. Where is the nearest plumber? How do I stop my child's nosebleed? What is the bus schedule for getting uptown? Whom can I call for information on how I can best help my friend with AIDS? How can I get information in the Vietnamese language on immigration rules? Did the mortgage company pay my property taxes? What's the high school class schedule today? Are there any part-time jobs in the area for my teenager? How can I renew my driver's license without going downtown and waiting in line for the whole afternoon? Can I leave a written message so the appliance repair service can find my apartment? What's the latest on the heavy weather warning for our area?

Now visualize how you can satisfy these needs with a device as easy to use as your telephone or TV set. You push a few buttons on a hand-held device, maybe speak few word commands; you hear or read the answers in the language of your choice, perhaps with added text or graphics on a small but very high resolution video screen.

Is visualizing a citizens' information service just so much "high-tech fluff" or the stuff of airline magazines? Not really. Like so many communications technologies of the past, it is possible to envisage how *existing* technologies can offer new services to society. Think for a moment how a few visionaries brought us new services that today we take for granted. Consider, for example, the vision of Samuel F. B. Morse of how a cranky railroad electrical signaling system of the early 19th century might be used to move important messages at the speed of light in the form of the "telegraph." Or consider the days early in the present century when Theodore Vail, the genius who shaped the American Telephone and Telegraph company into the world's largest corporation, imagined that a mainly business instrument like the telephone might eventually be found in every home. Add to this the image of early radio—mainly a ship-to-shore communications medium—as a new means for delivering music, announcements, news, and eventually advertisements to everyday citizens. Then there is television about which the April 8, 1927 edition of *The New York Times* reported "Commercial Use in Doubt." Given some thought about the images of old media when they were new, let us ponder the coming of electronic information services.

Over the past 20 years the coalescence of computing and telecommunications technologies has made possible the diffusion of powerful information, message, and transaction services available to anyone who can tap into the national telephone network. Increasingly, computers have come to handle the switching, information storage and retrieval, and our entering dialing commands into the telephone network. Simultaneously, the distributional power of telecommunications has brought computer-based capabilities to the fingertips of almost all telephone users. In one metaphor, it is as if the powers of computers, like the proverbial genie, have escaped from their dull gray boxes to diffuse throughout the everyday telephone network so as to be summoned up by a few flicks of our fingers on our phone push buttons or personal computer keyboard. To those of us whose first encounters were with computers that took up the space of an entire room, there has also been another transformation. The use of computers for creation, transfer, or manipulation of messages and images is far surpassing their original uses for numerical calculation. Just as writing allowed the extension of human vocal capabilities across distance and time, or the printing press greatly expanded the reach of the written word, the computer and its diffusion of powers via telecommunications is the latest in the evolution of media technologies.

Some will say, however: "What new medium? For those with access to computers, sending computer text messages over telecommunications lines has been around for over 30 years; newspapers experimented with and lost money in the last decade with a myriad of videotex services; the French have their touted Minitel system; and haven't you ever heard of Prodigy, Compuserve, Genie, or the coming of the National Research and Education Network?" Yes, these are all examples of the existence of forms of electronic information services but for the most part they are best characterized as "forays" into the public market rather than the likely form of an easily accessible, inexpensive, valuable, medium that a national policy could promote as universally available as with the telephone, radio, or television. You had to have access to mainframe computing to send electronic mail 20 years ago. Videotex was too expensive and duplicated too many other sources of information to be of much value 10 years ago. You need to be "into" personal computing if you are to be a typical user of Prodigy or Compuserve today. The blank stare you'll probably receive if you ask your neighborhood school about the National Research and Education Network is evidence that advanced telecommunications have not arrived truly on the local level. And if you think back to the young women in the opening anecdote of this chapter, you'll probably agree that when you do hear about everyday people using information services, their names are more apt to be Kimberly than Milagros. Their uses of service are more apt to be about stock quotes, games, and electronic mail than about information essential to everyday life in an increasingly complex, technological, and multicultural society. For the broad *public* at least, we have heard about network information services relative to the failure of videotex in America. The stories have been more about the ambitions of information providers than about people, more about markets than services, and—unlike America—more about regulation than free enterprise.

The key challenge, and the topic pursued in the chapters of this volume is that our federal government should seriously consider a national policy for promoting electronic information services as a new national medium available to all citizens as a "universal service at affordable cost," as we have already done for a half century for the telephone. This is a social policy proposal as well as a technological and industrial policy initiative for our times.

Our period right now in the early 1990s—at the time we are writing this book—appears most optimum for the vast expansion of electronic information services as a new public medium. For one, we can reap the benefits of major technological breakthroughs. The combination of three technological enhancements can provide the nation's telecommunications network with such a quantum leap in capabilities that we can have a virtually new public communications medium at our disposal. These enhancements are a low priced communications instrument with visual display and touch-screen input capabilities, the capability of "packing" multiple communications channels and thus mixes of voice, text, and graphics in a single telecommunications circuit connection, and the coming of powerful automated information storage, "packaging," and retrieval systems so that the network is virtually a publications medium. This leap could come in two stages—first, a mix of interactive voice, text, and graphics services, then as capacity enhancements allow, a full-scale interactive video communications medium. Because these services can be offered upon the existing "platform" of the nation's public telephone network—possibly combined with cable television—it deserves special attention as a powerful purveyor of "essential" information services for everyday citizens. Just as legislation leading up to and including the Communications Act of 1934 established broadcasting and telephony as vital to the public's interest, it is important to consider whether network information services are sufficiently important to the public well-being that we establish policy incorporating concepts like "the public interest, convenience and necessity," or "widely available service at affordable cost," again to recall some of the well known concepts of national communication policy.

That there is a potentially vast and lucrative market for a broadly deployed array of citizen information services is especially evident today in the vicious debates and political lobbying of newspaper and allied interests at preventing the nation's telephone companies from capturing this business. This debate was unleashed with the breakup of the nation's monopoly telephone company, AT&T, in a 1982 court decision that barred the newly divested regional Bell telephone companies from offering information services and the remaining downsized AT&T company from the business for 7 years from the implementation of the breakup in 1984. Mainly the argument is that telephone companies, especially those that continue to have a franchise to provide local service, should not be allowed into a business where they would have a monopoly advantage over competitors, or worse yet, where they could subsidize the new business from revenues guaranteed from their telephone franchise. The counter argument is that newspapers are mainly trying to block telephone company entrance into information services because they cannot stand new competition, especially for advertising revenues.

This standoff took a new turn in 1991, when the federal

judge who oversaw the breakup of the Bell system 7 years ago lifted the ban he had imposed on the divested Bell telephone companies entering the information services business. This meant that we now had seven companies whose combined annual revenues exceed some $85 billion that would in all likelihood become information providers, or as some say, "electronic publishers," in the decade of the 1990s. But this decision, however, has not left the telephone companies free to go into the public information business. Pro-newspaper interests are aggressively promoting both Federal and state level legislation to continue to bar telephone companies from being "content owners" in the new business—at least until it can be shown that they will not dominate the competition or cross-subsidize themselves from monopoly-protected revenues. On the other front, telephone companies are forming new unregulated subsidiaries to enter the information services markets and are gaining some compromise with newspaper and other competition interests in state-level agreements.

The outcome of all this? Many experts—and some in the chapters of this volume—see an eventual partnership among the most farsighted newspaper and telephone interests—mainly they will go in the information business together as partners. Just as it is unlikely that newspaper interests can continue to bar an $85 billion industry from going into new businesses in an age of newly charged economic expansion in this country, it is also unlikely that the information services business will grow without influence from an industry with nearly 500 years of tradition in serving the news and information needs of publics. (In chapter 2, we look at the newspapers in an electronic world of the future.)

Moreover, as of this writing there appears to be the final major boost to the establishment of new telecommunications services for the U.S. public. In 1992, we elected a president whose technical policy includes expanding new telecommunications services beyond the realm of research universities and big business to the everyday world of the small business owner, the local school district, and eventually to our homes. His vice president, Albert Gore, Jr., as a senator, sponsored a bill to build a national "information highway." In the context of this political shift of priorities for U.S. industrial policy, we may now see the emergence of a national policy for citizen information services—especially services for all citizens, not just a privileged few.

TOWARD A VISION OF A NATIONAL INFORMATION SERVICE

Let us next consider a national information service in a manner to make more concrete—or to create more of a vision—of what is at stake here.

What is a picture of a public network information service? Rather than describe it in terms of abstract policy or goals, what would it look or feel like? Mostly, the picture of videotex has conjured up visions of young, upscale moderns clicking their computer "mouses" on color screen menus, checking the stock market, or doing "on-line" shopping. A much more practical and potentially universal information service could look much less dramatic—at least in its first incarnation. It could also be more directed, like basic telephone service, at the broad spectrum of the U.S. population, including the poor as well as the middle class and wealthy, the rural dweller as well as the urbanite, and the elderly, disabled, and the newly arrived to fill out that spectrum.

Again, imagine a hand-held communications device, perhaps a next-generation common telephone, similar to those that now occupy our residences save for the presence of a crisp visual display panel, about 3 x 5 inches in area. Figure 1.1 gives a visual "feel" for this phone, which is on the planning boards of both domestic and foreign manufacturers already. This telephone is a step ahead of today's voice-only model, just as touch-tone and a few extra special purpose buttons replaced the rotary dial. You will be able to talk on this phone just as you probably have already done today, except that as it is useful, you will also simultaneously be able to send and receive information via the visual screen which at your command can be configured into a custom ("touch-screen") keypad. You could, for example, call up a screen for school district information, including vacation schedules, announcements, sports events, dates and agendas of board meetings, budget summaries, and perhaps even the record of your youngster's absences and tardies.

FIG. 1.1. Examples of telephones designed for information services.

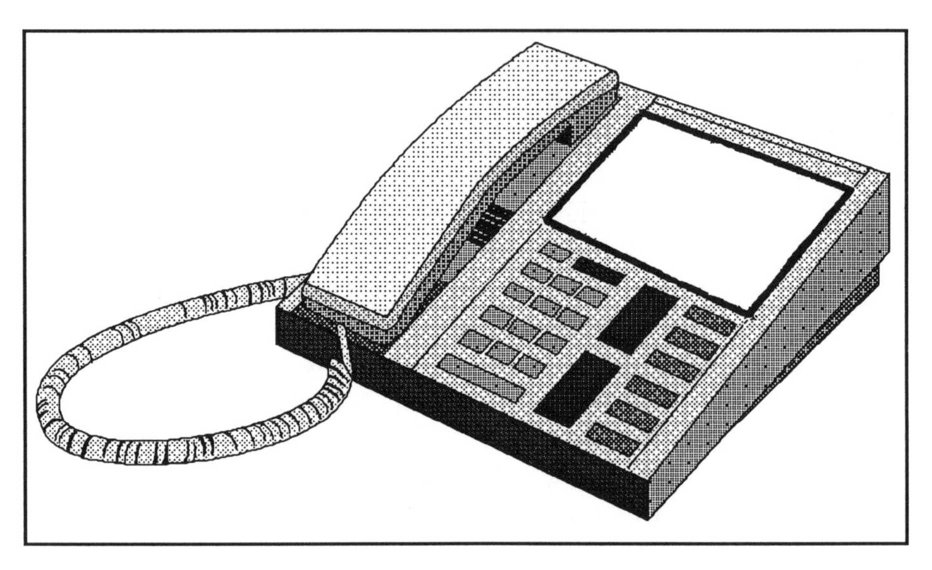

Nothing too fancy here. Experience shows that if this phone were any more complicated than operating an automatic teller machine or your television set, or maybe the newer, more easily used videocassette machines, it would fail to be used. Among the practical criteria for its use is that you should be able to perform routine functions easier, faster, or more accurately than you can now do by voice alone.

Imagine that you were to use your new phone for directory assistance in another state. You would have a national telephone directory at your fingertips. Beginning with any information you had on hand, a city, an area code, you could work your way down to directory entries that fit your request. Touch the entry of your choice and the number is automatically dialed. (An electronic telephone directory was one of the main functions of the French national videotex service called "Minitel.") Without taking the space to describe them, many other "housekeeping" functions of telephone use (e.g., call forwarding, caller identification, or to block your own identification if desired, checking the balance of your current charges, and so on) could be performed with combinations of text and voice.

Many of your routine uses of the phone you now have at home could be more efficiently accomplished. Emergency calls for police, fire, or medical assistance could be a matter of one or two touches of the screen, with new provisions for verifications to alleviate false alarms. Vital information on the location and any special conditions would be immediately available to the emergency service providers. Simple information-seeking or transaction uses of the telephone would be enhanced. Applications ranging from checking transportation schedules to reviewing bank balances could be semi-automated by use of the text and numeric input capabilities of the screen, including configurations of menus for selecting alternative information sources. The data-handling capabilities of the system could also support automatic meter reading of utilities such as water, gas, and electricity. As with some current services, you could have the option of using the telephone system to transfer funds from your bank for bill payment.

Social use of the telephone can also be enhanced by the ability to share text as well as voice messages, to combine the two, and provide a list of callers by their numbers and names. You could also leave special text messages to be sent when specific numbers call in, a more or less customized answering or message service.

But you might already be thinking that we can accomplish many of the foregoing services now in one way or another. Up to this point we have painted a picture of a slightly enhanced, "video-augmented," telephone service. The real advance comes in the near infinite variety of information that can be made publicly available via this system. This information could be of a public service nature, such as emergency medical advice, or more commercial such as electronic *Yellow Pages* ™, or retail catalogs from which you can order directly over the network.

The vision of a national information service does not have to hinge on an upgraded telephone instrument; in fact, some experts, such as Roger Fidler (chapter 2), make an interesting case that new types of screening devices of a newspaper nature will best serve our age of electronic information services. Fidler envisions a light-weight, cordless, flat screen panel that could be held as easily as today's newspaper while you relax in your favorite chair while scanning the news or looking up specialized information.

Now it is necessary to assume that every citizen would need to have a private link into the service. There could be a major program to develop public information kiosks available in public spaces (see chapter 5).

ESSENTIAL INFORMATION AS A "UNIVERSAL SERVICE"

Of high public service priority is to offer the information or information-exchange services most essential to the citizen's well-being in our increasingly complex society. What would most benefit citizens and, in turn, society? Beyond the emergency services just mentioned is information mostly of a human service nature that lends itself well to brief text or graphic displays and "branching" menus so one can either get to a needed service quickly or else browse to study available alternatives. Table 1.1 summarizes examples of services many would consider of an essential nature. Access to this information would be a part of "basic" service, the costs included as a

TABLE 1.1
Essential Information Services

Essential Information Services
Abortion Rights and Counseling
Alcohol Abuse
Bilingual Instruction Services Guide
Bus or Subway Schedules and Current Conditions
Calendar of Local Events
City and County Offices Services Guide
Child Abuse
Community Services and Agencies
Consumer Warnings
Credit Services and Rights
Directory of Telephone Numbers
Divorce Laws and Counseling
Drug Abuse
Environmental Hazards (Weather, Toxins)
Financial Information
Hospital Guide
Immigration Laws and Rights
Insurance Terms and Tips
Job Listings
Legal Terms
Library Services and Card Catalog
Literacy Instruction Services Guide
Local Transportation Guide
Medical Emergency Tips
Nutrition Guide
Pregnancy Counseling
Rape Counseling
School District Calendar and Current Public Notices
Spouse Abuse
Utilities Information and Access Numbers
Weather, Local
Welfare Rights

flat rate in the basic phone bill. In effect, this is a proposed upgrade of the traditional definition of "universal" service.

In many cases, essential information could be contributed by local government, public service agencies, or sponsored by commercial interests. Accessing these information services should be an efficient combination of voice, text, and graphics exchange between you and the upgraded telephone device.

Until speech recognition technology advances further, speech will be mainly one-way as synthesized by the system in the form of messages to you (and here you could indicate the language of your choice).

Use of the system should be tailored to the nature of the information needed. Treating a nose-bleed should only require a "touch" or two in order to get down to precise directions, including graphic illustrations, for emergency treatment. By contrast, how to examine a persistent red spot on your arm as possibly cancerous could provide more information on different characteristics and types of cancers, including where to get help. Information on pregnancy and even some of the counseling could be provided by the network; it could be private, anonymous, and in one of the major languages spoken in the client's community. Referrals for where to get in-person counseling —so sadly lacking in the opening vignette—as well as travel directions and scheduling could be offered by the network. Quickly accessed bits of information might best be communicated by text and graphics, whereas counseling information could be recorded in a reassuring voice. Furthermore—and this is especially important regarding emergencies—voice messages and screen icons could transcend the barrier of illiteracy.

Beyond the "essential" information services, there would be a much wider range of commercial services. They would be offered by business sponsors either as advertising or for a user fee, or both. Some of these have already been advocated in various consumer surveys, or found to be popular in current on-line services, including "audiotex" and "900" number services. We return to the topic of commercial information services at a later point.

WHY AN EXPANSION OF THE TELEPHONE NETWORK?

How could an enhanced telephone receive so much at once over common telephone lines? This is the second of the technological advances introduced earlier. As you may know, the public telephone network was originally designed to transmit simultaneously in either direction a basically intelligible voice message and no more. This is a very limited "pipeline" when one considers all of the additional message traffic we wish to transmit—from high fidelity voice and music, to data or images.

One technological solution is to increase the physical capacity of the line itself, as in replacing the "copper pair" with coaxial cable or fiber optic lines. But another, and very practical approach, has been to compress the signal by digital coding in ways that allows for much more to be carried by traditional copper lines. Currently the best known application of this type is called integrated services digital network (ISDN for short) which combines two higher than typical capacity voice or data channels with one slower signaling channel, importantly all on the same "copper pair." (Plainly speaking, it "stuffs" more messages in the same pipe.) With ISDN service you could observe a shared and changing text and graphics display screen while talking at the same time with your connected party. A higher resolution and faster moving image display could be supported by a combination of the two main ISDN channels while the signaling channel communicates your "dialed in" choices. Especially important is that the basic form of ISDN does not require replacement of existing voice grade lines in the network. But it does require new switching equipment as well as special interfaces between customers' equipment (telephone, terminal, computer, fax, etc.) and the public network. Your newly enhanced telephone would come with that interface. Although ISDN has been touted mostly for business applications, there are persuasive arguments that it would be an excellent "platform" for the first stage of a public information network.

We must note, however, that not all stakeholders in our nation's telecommunications scene favor ISDN. The cautious, such as consumer advocates who fight any increase in public communications costs, claim that we simply do not need it. Many telecommunications optimists counter-claim that ISDN is not enough to meet our 21st-century needs, and argue in favor of "leap-frogging" this technology in favor of going directly to building a much higher capacity ("broadband") public network, a point to which we return. One point to concede is that the network eventually will likely be broadband so there is a genuine issue of whether the ISDN route is a logical intermediate step. In fact, Bellcore (the research center of the Bell operating companies) president, George Heilmeier, argues that the Bell companies will probably move directly to a more advanced, broadband network, or as they call it, the advanced intelligent network (see AIN; chapter 4).

It is important not to overlook, either, the potential for cable television components in a national information service. If, indeed, the service will involve a broadband link into every home, why not patch into the existing cable television systems of the country?

A MARKETPLACE FOR COMMERCIAL INFORMATION SERVICES

And now back to the information services. We can add to our list those services that although important to many are not truly essential. These have been determined in various consumer surveys, or found to be popular in current online services, including "audiotex" and "900" number services. These could be offered by commercial sponsors, for a small user fee, or both. These "optional" services are included in Table 1.2.

TABLE 1.2
Optional and Commercial Information Services

Airport Schedules and Current Conditions
Classified Advertising
Home and Garden
Horoscopes
Insurance Terms and Tips
Investment Information
Mortgage Facts and Rates
Movie Reviews
News Headlines
Recreational Attractions
Ski Conditions
Soap Opera Updates
Sports Events Schedules
Sports Scores
Television Program Schedule
Weather, National or International by Location
"Yellow Pages"™
Zip Code Directory

Imagine the value of a public directory of home services. You have a dead electric circuit. If you did not choose to use the service as a guide to home repairs, you might go directly to the "electrician's" icon on a home service directory screen. The information service, knowing the location of your telephone, could list electricians according to their proximity to your address.

If the information services are to succeed, the system itself must have a built-in capability of "finding the market." Choices among services must be highly flexible, especially for those of a commercial nature. It should not be much more difficult to be an "information provider" than, say, to sell your goods at a local flea market (an analogy we pursue in a moment). Since modern network systems can easily track frequency of use and maintain billing records, those services seldom used can be omitted or modified. The service could also have an option so users can submit suggestions for new entries.

This flexibility can be provided by the third technological advancement mentioned earlier—namely, computer-based capabilities for information receiving, storage, "packaging," retrieving, and billing. Such systems can facilitate easy entry of information by providers, public service as well as commercial. Information on local transportation schedules might be routinely entered by the municipal transportation authority, perhaps from their own electronic data files. Electronic *Yellow Pages*™ could be regularly maintained by the local exchange telephone company. It should also be easy for information providers to submit, modify or remove their entries.

Again, you or I should be able to try our hand at being information providers. Perhaps we have valuable experience in the recycling of paper materials for garden or plant mulch. A truly public national electronic publishing system would make it an easy matter for us to offer this information for a small fee. We would enter our information as a "publication" in the national electronic service. Any customer who uses our recycling information service is billed for the purchase; a small part of the fee would go to the information service operator and the remainder would be paid to us as the "information provider." In the long run, if our information-provider income exceeds costs, we are now in the information business. But if nobody needs our information on recycling or won't pay the prices, then the market has spoken and we, like many other fledgling information or service providers, may go on to try something else. There could also be public service versions of information providers. Perhaps we are an agency charged with disseminating information on the prevention of teenage pregnancy. We have a budget for information dissemination, and we may choose to make expenditures to post our information free of charge to any users who may wish it. We pay the information service provider to "publish" our materials but do not take a fee from users. Such public service applications may be an important catalyst for development of large scale information exchange because there are already funds being expended for information dissemination that could be invested in the new system.

To continue with a publishing analogy, we might envisage the system as one giant "electronic flea market." Customers have easy access and so do vendors. Importantly, many new and innovative services can easily be market-tested. This marketplace could open up vast new possibilities for small entrepreneurs heretofore nonexistent in publishing, or telecommunications for that matter. This also avoids one of the greatest shortcomings of earlier videotex trials: inflexibility of offerings. Large publishers appeared very much intent on seeing their services as an extension of the print medium. When electronic mail and "chatting" began to grow, especially with sexual overtones, one major publisher made it clear (in remarks at a conference) that he had no intention of operating "electronic services for the lovelorn."

The concept of an automated "electronic publishing" is not a pipe dream. It exists in various smaller scale forms already. But probably most important is that a system of this type is among the major projects of Bellcore, the research organization jointly supported by the Bell companies.

WILL CITIZENS' INFORMATION SERVICES FAIL LIKE VIDEOTEX?

Despite the opportunities presented by the coming of a new medium of public information services, there remain many naysayers. One argument is if the millions invested by publishers in videotex did not "make it," why should the next venture do any better? On the other hand, to take a positive tack, what lessons can be learned from failures of mostly the publishing industry in this area?

In one respect, the very basic examples given thus far are meant to make the point that most previous ventures into electronic information services have overlooked two important possibilities. First, most videotex ventures did not begin with an emphasis upon what the broad public most needs and would likely use from an information service. They went for the "upscale" markets, the people who could afford expensive terminals, pay for special services, or (later) who wanted to expand the uses of their personal computers. They also made the mistakes that many have summarized as follows:

- Publishers often offered services that were available easier or less expensively in other forms. (I once subscribed to an expensive online version of the *Wall Street Journal* only to find that it never contained information any earlier than the paper copy that arrived on my Los Angeles doorstep at 4 in the morning.)
- Providers were inflexible in offering what users tended to gravitate toward in their continued uses of the service—as mentioned earlier, spending more time with electronic mail than in retrieving news features.
- Many videotex ventures failed to offer a single compelling service, something that would "trigger" a customer's continuing subscription (as so well documented in the book *Electronic Publishing Plus* edited by computer expert Martin Greenberger).
- Many ventures that started with low rates, "lent" terminals, or trial use periods, died quickly when customers were asked to pay a price for the service that was realistic in terms of production and transmission costs, but not value assigned by customers. Most of the videotex trials never grew large enough to achieve the economies of scale associated with most public or mass media.

Another negative argument refers to the few trials that telephone companies have already made into videotex, but where they conveyed the information services of others. They were mainly the "conduit" and the "billing agent." Although at first Judge Greene barred local telephone companies from being involved in any information services, he relented several years ago to allow them to be "conduits" but not "owners" of the information they provided. A number of trials ensued, and most were unsuccessful mainly for the same reasons as the publishers (phone companies did their homework poorly), and also because they lacked control over the materials. Phone companies quickly learned that "conduits" do not make much money in this business, although they are very much the target of public criticism over services that may exploit children, or that allow unwitting customers to run up big bills. If they did not learn this in their brief encounters with videotex partners, they are learning it now from "900" number services.

But on close examination, what the telephone companies lacked was the ability to engage in a winning "trigger" service, namely electronic *Yellow Pages*™. This is the service that publishers tend to fear most although their campaign against telephone-based information services has included arguments that phone companies could unfairly subsidize information services from their regulated businesses. Publishers have also decried the possibility of an "Orwellian" world where "conduits control content." But mainly, publishers simply do not want to lose advertising revenue to a new medium. (Read the Newspaper Association of America's *Presstime* if you want to learn every possible negative reason for barring phone companies from the information business. Its stridency has worn thin as of late.)

Ironically, persuasive arguments can be made that for other than classified advertising, newspaper publishers do not have that much to lose to telephone companies, and even that loss is debatable. There is no evidence that customers prefer a screen to the ease of scanning a well designed printed page. There is evidence that newspaper-type "hard copy" is much more conveniently delivered than possibly printed out in the home. After all, who wants to "read their telephone" at the breakfast table! It seems likely that the publishers who have broken new ground in the last decade—those who have been visibly innovative—will benefit by capitalizing on partnerships as well as product or service contrasts with network information providers.

Telephone companies have quite limited experience in the marketplace of public information and entertainment; they need publishers and producers as partners. Selling content is a whole different game from conduit. (Can you imagine an engineer and utility executive of some 30 years doing a Beverly Hills media deal with the likes of Rupert Murdoch or Barry Diller?)

THE TIME FOR DECISION

Quite obviously, the issues discussed in this chapter have, and can, go on and on. But the fact remains that the mid-1990s is likely to be a pivotal milestone in this country's position on the large-scale entry into network-based public information services. Technologically, a new medium exists but it has been restrained by regulatory forces, if not also a lack of imagination and research of those who have ventured early into the videotex services. Clearly, the concept of a public telecommunications infrastructure—like universal connection to the network—may be giving way to the concept of access to an *information* infrastructure that sits upon the platform provided by the coalescence of computing and telecommunications. Large businesses have already discovered the value of this resource. Japan and the countries of the European Economic Community have declared the development of an information infrastructure a part of their social investment for the 21st century. Hopefully, we in this country can recognize it as a priority for our citizens—<u>all</u> of them, including those who may most need it—like ...

.... *las amigas de Milagros.*

ACKNOWLEDGMENT: This chapter is adapted from Williams, F. (1991a).

Chapter 2

NEWSPAPERS IN THE ELECTRONIC AGE

BY ROGER FIDLER

Will newspapers fulfill citizen information needs in the electronic age? Only out of an ignorance of media history would one see the future of electronic information services mainly as a newspapers versus telephone companies contest. History shows that new media do not typically replace existing media, but instead modify them. Newspapers are already well into the electronic revolution. They are largely written, edited, transmitted (where necessary) to printing plants, and archived electronically. The only remaining tentative or experimental step in adoption of electronic technologies is the final delivery link from the publisher to reader. It would be shortsighted to consider the prospects of a national information service without carefully examining the likely role of newspapers. After all, they represent three centuries of experience in the delivery of information to the public.

What, then, is the future of the newspaper in the world of electronic information services, or better yet, what is the outlook for the electronic *newspaper? Few could pursue an answer better to this question than Roger Fidler. Roger had first-hand experience in videotex experiments of the early 1980s and now serves as director of the Knight-Ridder Information Design Laboratory, recently founded in Boulder, Colorado, as well as director of new media development for Knight-Ridder Newspapers. He was a 1991-92 fellow at The Freedom Forum Media Studies Center where his research involved forecasting the future of newspapers as described in his forthcoming book,* Mediamorphosis: The Transformation of Mass Media. *This chapter draws from portions of his book and presentations he has made on this topic.*

VISIONS OF THE FUTURE

This century has seen the emergence of more new forms of communication than in any other period in history. Instead of discarding the older forms, we have simply added to them. With each new medium, our choices for information and entertainment have been greatly expanded. While this may be seen as beneficial for individuals, the consequence for mass media companies has been an ever-increasing fragmentation of audience and advertising support. Because this trend has been consistent for so long, most of today's media experts are predicting that we will see more of the same in the next century—more media, more information, more choices, and more fragmentation. If we extend this process to its logical conclusion, then we should expect all forms of mass media to disintegrate, within a few decades, into a myriad of low-budget, narrowcast programs and intensely personal newsletters. Other, more technological pundits arrive at a similar conclusion, but by a somewhat different path. They are inclined to see all of us as dissatisfied information seekers, hungering for more facts at our fingertips. For them, all contemporary forms of mass media—newspapers, magazines, television, and radio—are inadequate relics of the past. To satisfy this perceived hunger, they see the future of human communications residing in omniscient databases and intensely personal news services delivered directly to our computer printers and fax machines.

Different as these views may be, both bear a striking similarity to those of traditional prognosticators. One group plays it safe by incrementally extending known trends and proven technologies, while the other leaps to a popularized "futuristic" technology. The flaw in these views is that they are based on the technologies and ideas of the past. Most of today's media experts are missing the emergence of what may be the most significant new media development since the invention of printing, the *flat panel*.

This technology, I believe, is the fundamental technological milestone underlying the future of the newspaper.

THE COMING OF "DIGITAL APPLIANCES"

Flat panels are best described as very thin picture tubes. They are considered the digital successors to the bulky analog computer monitors and television screens that consume so much space on our desks and in our homes. The light-weight portable computers now so commonly used by business and professional travelers would not have been possible without flat panels. However, the most important uses for flat panels may not be as mere replacements for picture tubes. Just as the solid-state transistor transformed electronics in the 1950s and 1960s, the flat panel has the power to transform communications, and with it all forms of mass media, in the next two decades. Flat panels combined with microprocessors, memory, and communication links are now beginning of take on an existence all their own. These devices, which are often generically referred to as "digital appliances," (see Fig. 2.1) represent an entirely new class of electronic media. Rather than a specific technology, digital appliances can employ a variety of technologies and can take many forms. Instead of keyboards, most will use electronic pens for entering and interacting with content and will be capable of recognizing handwriting and voice commands.

In the next few years we will see a plethora of portable digital appliances, such as the recently introduced *Apple Newton* and *EO Tablet*. While these devices may seem insignificant at the onset, they represent the first wave of products and services that are likely to emerge in the next two decades from the implosive convergence of computer, communication and information technologies. By the end of this decade an array of lighter weight and far more powerful flat-panel devices should begin transforming the promise of multimedia publishing into commercial reality. They will range from

Figure 2.1
Digital Appliance

Dow drops 2.94 points to 3,282.41

Crichton's "Rising Sun" a best seller

A novel way to make screens flat

The New York Current

WEDNESDAY, FEBRUARY 17, 1993, 11:47 PM MDT

U.N. commander halting attempts to reopen airport

SARAJEVO (06-21-92 0612 EDT) — The United Nations announced today it was ceasing all attempts to reopen the Sarajevo airport until there was a cease-fire for at least 48 hours. The U.N. commander, Maj. Gen. Lewis MacKenzie, acknowledged with obvious anger that it was possible the U.N. mission would fail.

Czechs and Slovaks sundered by politics

PRAGUE (09-15-92 01544 EDT) — With the demise of communism behind them, Czechs and Slovaks are looking into a future divided. Although the national union between the peoples prevailed for 74 years, Czechoslovakia, like other countries that were bound to the Soviet Union, is reassessing the economic and political shifts of the post-cold-war era. Differences of leadership, opinion and vision have already separated the two peoples, and there is talk of making the division formal.

Basic services gone, Kauai tallies hurricane damage

HONOLULU (9-15-92 1800 EDT) Children play in rubble left by the fury of Hurricane Iniki Monday on the island of Kauai. Limited phone service has been restored for some of Kauai's 52,000 residents. About half the island has running water, and electrical power is expected to be restored to the largest town on Lihue by next Monday.

Rates cut provokes worldwide stock rally

LONDON (6-15-92 1900 EDT) — The announcement by Germany's powerful central bank that it would make cuts in its interest rates has sparked gains in all stock markets. But the optimism surrounding them may evaporate if France rejects unification.

Satellite gap limits views of hurricanes

The United States is facing heightened danger from hurricanes because delays in modernizing its weather satellites are forcing the use of spacecraft so antiquated and poorly positioned that they are raising the risk of forecasting errors.

Doctor's group offers plan to curb costs

WASHINGTON (9-15-92 1800 EDT) — The American College of Physicians, the largest medical specialty group in the United States, said today that there should be an overall national limit on health-care spending, with new restrictions on the amounts charged by doctors and hospitals. The proposal signals a turning point in the debate over the future of the American health-care system because it indicates that significant numbers of doctors now support an approach favored by many labor unions and big businesses.

Bill Clinton greets Ladybird Johnson during a recent campaign stop at the University of Texas in Austin.

Peruvian television shows rebel leader in his cell in Lima.

Peru's rebel movement is still seen as potent

LIMA, Peru (9-15-92 1930 EDT) — With the arrest of Abimael Guzman Reynoso and at least three other senior leaders of the Shining Path, much of the rebel movement's upper echelon has been captured or killed. But experts warn that there could be years of continued violence before the insurgency is quelled.

Showdown with Iraq expected later this week

WASHINGTON (09-15-92 1543 EDT) – An impatient United Nations Security Council has demanded that Iraq comply "fully and unconditionally" with terms of the Persian Gulf War cease-fire resolutions by March 21, or face serious but unspecified consequences. As usual, it is unclear what Saddam intends to do. Deputy Prime Minister Tariq Aziz, was at first defiant but later he seemed to be seeking a way to back down as Iraq has done before.

Menu | Profile | Extras | Find | Save | Print | Quit

TODAY 2

Summary
General
Metro
Region
Nation
World
Lifestyle
Obituaries
Opinion
People
Weather
Ad Index

BUSINESS

GUIDE

SCI/TECH

WORLD

"smart" flat-screen televisions to light-weight tablets, the size and weight of a standard magazine, that can be used while lying in bed, riding on a subway, or sitting on a park bench. Ultimately, these devices can be expected to combine the readability and ease of using paper with the interactivity of personal computers and the compelling qualities of video and sound.

Portable digital appliances are likely to have a vast number of practical applications early in the 21st century. They will be used by factory workers as electronic clipboards and manuals, by executives for viewing and distributing memos and reports, by salespersons for presentations and order entry, by accountants to prepare tax returns and update spreadsheets, by repair persons and installers to access up-to-the-minute schematics and animated instructions, by public speakers as portable prompters, by students and teachers to access current editions of interactive textbooks and reference materials, and in nearly every other situation where paper is used today for storing, displaying, capturing and distributing information that requires frequent updating or is of an ephemeral nature. Into this last category we can include newspapers, along with magazines, journals, reports, and most books.

The flat panel will, in essence, offer a reading page, and would allow you to use a pen for navigation. You would still have the ability to annotate information, to highlight and circle material, to save or print copies. My assumption is that we would print much less than we do today because the characteristics of the information displayed on these tablets will be much more comparable to ink on paper.

However, the significance of portable digital appliances is not just in the creation of new high-tech gadgets and electronic wizardry. It is in their embodiment of a concept known as "media convergence" that their greatest importance is found. This has many implications for the future for the electronic newspaper.

MEDIA CONVERGENCE

Until now, we have tended to see newspapers, magazines, books, broadcast television, radio, satellite cable systems, telephones, and personal computers as distinctly different forms of communication and information media. But they may not be as different as they seem, particularly in their roles as providers of general news and information. From the perspective of consumers and advertisers, all forms of media are seen as components of a single information system. The distinctions of sound, motion and print are incidental. Advertisers as well as customers have always tended to allocate their time and money to the media mix that fulfills their needs in the most convenient and economical ways.

Even within each form of mass media, the distinctions are rapidly becoming blurred. Newspaper photographers are beginning to use video cameras on assignments and are routinely capturing still images from television news programs for reproduction in print. Time/Warner, ABC and other media companies are distributing video clips combined with textual information on interactive video discs and CD-ROMs. At the Knight-Ridder/Chicago Tribune Graphics service offices in Washington, D.C., newspaper artists are using Macintosh computers to create animated graphics for television as well as still graphics for print.

What this blurring of distinctions represents is an early stage in the process known as "media convergence." In recent years, the words *convergence* and *multimedia* (the result of media convergence) have been commonly used as catch-alls to describe any situation where two or more forms of media are blended or used in conjunction with each other. But that describes only the tip of the new media iceberg that is about to collide with our contemporary perceptions of mass communication.

The concept of *media convergence* can be traced to the minds of a few visionaries back in the late 1960s, but it wasn't until 1979 that it was given a visual signature in the form of three overlapping circles labeled "Broadcast & Motion Picture Industry," "Print & Publishing Industry," and "Computer Industry." That diagram was drawn by Nicholas Negroponte as a marketing symbol for his proposal to build a Media Laboratory at the Massachusetts Institute of Technology in Cambridge. Negroponte and others at MIT foresaw the coming together of these three industries to form a new communication model. While they believed each industry would continue to evolve, the area of richest opportunity was seen at their intersections. In Stewart Brand's (1987) popular book, *The Media Lab,* Negroponte is credited with recognizing that all communication technologies were suffering from a joint metamorphosis, which could only be understood properly if treated as a single subject, and only advanced properly if treated as a single craft.

Negroponte and former MIT President Jerome Wiesner used this argument to successfully raise the millions of dollars needed to build the Media Lab in 1985. Under Negroponte's direction, the Media Lab has pursued several "opportunities at the intersections" including electronic newspapers and digital high-definition television (HDTV), but it is in the area of interactive multimedia publishing that its research had the most immediate impact. Although the Media Lab was not the first institution to become involved in multimedia research and development, it succeeded where others had not in focusing attention on the potential of multimedia publishing and played a major role in popularizing the concept of media convergence.

Today, dozens of companies, large and small, are developing interactive multimedia applications for personal computers ranging from children's books and educational games to historical documentaries. However, until recently, the high costs for the computer equipment needed to run these applications prevented them from reaching large audiences.

In an effort to make multimedia applications more accessible, major consumer electronics companies, such as Sony, IBM, Philips, and Apple Computer, are beginning to market relatively inexpensive devices that integrate CD-ROM players, similar to the CD players used for music recordings, with standard home televisions and, in some cases, with small flat-panel displays. If portability becomes as important to multimedia systems as it has for audio systems, these new devices could help drive the development of low-cost, light-weight panels for other digital appliances. And in turn, low-cost, portable digital appliances could help drive multimedia applications for mass audiences.

There can be no doubt that media convergence will have profound implications for all of us in the coming decades. The physical blending of print, video, sound and computing within emerging digital appliances clearly represents a fundamental shift in our methods of communicating and interacting with information. In the ways that mechanical printing presses and ink-on-paper publishing have defined the present era, it now appears likely that personal electronic "presses" and multimedia publishing will define the new.

WHAT HAPPENED TO THE PAPERLESS SOCIETY?

If you are beginning to feel a bit skeptical about the future as I've painted it, that is understandable. For more than 20 years, pundits have been predicting the coming of the Information Age and the "paperless" society. But until now, the visible effects of this revolution on society as a whole have been largely confined to office automation and satellite-delivered cable television. Personal computers, which were originally seen as the heralds of this new era, are still used by most people as little more than electronic typewriters. Even with new "user-friendly" software and the addition of a "mouse," personal computers remain decidedly "unfriendly" to the majority of office and factory workers, especially executives.

One of the problems we face in trying to conceptualize paperless publications is that personal computers have proved to be such poor substitutes for ink-on-paper communication. Instead of less paper, as futurists predicted in the late 1970s, we are now dealing with more paper than ever. Although personal computers have had a measure of success in making the creation and revision of documents more efficient, they have clearly failed as tools for sharing and interacting with information. Given the choice between reading information on a computer monitor or reading it on paper, most people will not hesitate to choose paper, especially when the content consists of general news and information or leisure reading material.

The reasons are numerous. Most computer monitors are set on desks in fixed locations and positions that are uncomfortable for easy reading, their displays are crude and hard on the eyes, and typographic presentations are generally monotonous and vastly inferior to print. With computers, comparing items that are found in different parts of a document or in several documents is difficult and time consuming. And navigating through seas of directories and cryptic titles is a frustrating experience for all but the most dedicated desktop computer users. Larger monitors have become more common in recent years, but our ability to browse large amounts of electronically stored information quickly is still severely limited. When it comes to sharing information, the exchange of computer documents via electronic networks is still fraught with frustrating and often insurmountable problems. Even in offices with sophisticated electronic mail systems, copiers, and fax machines remain the tools of choice for distributing documents that have been created on personal computers. Moreover, for most of today's adults, personal computers are seen as tools for work, not pleasure, which is one of the principal reasons why visions of electronic newspapers, magazines, and books have gained so little popular support. Even among those people who design and program computers, the idea of doing their leisure reading on a computer screen generates little enthusiasm.

All of this would seem to indicate that newspapers and other forms of print media are just fine as they are, and that no electronic form of "print" media has a chance as a competitor anytime soon. That is the conclusion publishers would most like to hear. But to accept such a conclusion could be perilous as well as shortsighted, for it is in the distinctions between personal computers and portable digital appliances that the opportunities and the threats are found. It is most important to recognize that the portable digital appliance is not a personal computer as IBM defined it in the 1980s. Rather than a tool for creating and revising documents, the PDA is primarily a tool for communicating with other people and interacting with information: more like ink-on-paper than a PC. In a very real sense, these devices will provide us with the ultimate recyclable communication and information media—*digital ink on silicon paper*.

With contemporary personal computers, we are constantly made aware of their underlying technologies, even when we are performing basic tasks. That is not the case with emerging digital appliances. Just as our use of traditional ink-on-paper communication no longer requires us to understand the underlying technologies of printing and paper manufacturing, computers and telecommunications are evolving toward a point where these technologies will be totally transparent, which is as it should be.

Another important difference is in the primary sources of information. Until now, most of the information stored in personal computers has had to be entered by the people who use them. With few exceptions, most of the information entered is formatted for traditional printing and reading on paper. The situation is likely to be reversed with portable digital appliances. Most of the information will come from other sources, and it will be formatted to be read on the screen. Copies may occasionally be printed on paper, but in most cases the screen images will suffice once the display quality matches that of ink on paper. Sharing annotated documents with others should become even easier than using a fax machine.

Of course, history shows it takes much more than clever technology to make a new communication medium successful. Portable digital appliances will have to be significantly better than paper, without sacrificing any of its benefits to become a popular alternative to ink on paper. To succeed as a mass medium, they will have to be more compelling and convenient than existing forms of mass media, and so accessible and intuitive that everyone can use them on the first try without a manual or any skills more complicated than turning pages or changing television channels. Even with these qualities, the portable digital appliance is unlikely to become a ubiquitous medium until it is perceived to be affordable. Many unforeseen events can affect the adoption of a new media technology by mass audiences. The adoption of television, for example, was delayed for more than two decades by the Great Depression and the Second World War. But barring such setbacks, most researchers agree that if the developments of flat panel displays and computer memory chips continue at their current pace, the basic criteria for adoption will be met early in the next century.

DELIVERING THE ELECTRONIC NEWSPAPER

Although portable flat-panel devices hold the most promise for displaying electronic newspapers, they would be useless for breaking news content without a fast and cheap method of delivery. For general audiences the cost for electronically delivered editions will have to be comparable to that of printed editions, or even less.

There are currently five ways that newspapers could be electronically delivered at low cost. One way is to use cable television channel. The cable "guard band," which is an unused area, can handle data transmission rates of up to 9,600 bits per second very effectively. However, this would be very slow for multimedia packages. Although there would be little cost associated with delivery on the guard band, a multimedia newspaper could take several hours.

Delivery could be accomplished overnight, but that is not going to be adequate for the long run.

Another way to do it would be to take over an entire cable channel. There is sufficient bandwidth in a cable channel to deliver literally hundreds of multimedia newspapers. Each newspaper could be bundled as a package, with each one taking a minute or two to deliver.

Your digital appliance would be set to select and capture the papers you subscribe to. There are a number of pricing models that could work here. One that looks most promising would be to have a small monthly fee for access to the "newspaper channel" collected by the cable operators. Each newspaper would then continue to deal directly with the customer, establishing a subscription price.

A second method for delivering the electronic newspaper that has been discussed for some time involves using fiber optics to the home. Fiber to the home still looks like it could be a decade or so away, if it develops at all, although one never knows with the regional telephone companies, and how aggressively they will move to introduce the fiber. Moreover, with compression, some of today's technologies make it possible to deliver a multimedia product to the home without having fiber installed.

A third alternative would be via broadcast television, particularly after we have the high-definition television (HDTV) bandwidth available.

A fourth way would be to use direct broadcast satellite (DBS), which Motorola and other companies are now working on. DBS is more uncertain, but it would provide a low cost method of delivery for the rural areas of the world, or places where communications are not adequate to handle multimedia delivery.

A fifth option that is being considered by book publishers is to use memory cards. With these credit-card sized storage units you could go to a bookstore vending machine or a newspaper kiosk and select the books or newspapers you wanted, load them onto the card and then insert it into your flat-panel device. Memory cards currently have a storage capacity of about 8 megabytes, but the expectation is that they will eventually get up to the range of 100 megabytes or even a gigabyte. This credit-card format for mass storage looks very promising as one way of providing electronic publications.

EVOLUTION OF THE ELECTRONIC NEWSPAPER

To help you visualize how portable digital appliances may be used in the next two decades, I have attempted as best I can in print to describe my own vision of electronic newspapers, as well as several other applications of multimedia publishing that may be common by 2010. As with most science fiction, the following scenario is based on current trends and the plausible projections of people who are intimately involved emerging technologies.

SEPTEMBER 21, 2010

As I awaken, the view from my hotel room window gives no clues to my location. All I can see in the darkness is a galaxy of street lights. My body is telling me it should be about 7 a.m., but the digital clock next to my bed argues that it's 4:17. Memory returns slowly. Finally, I recall this is Eugene, Oregon, and that I arrived late last night from Miami to speak at a conference this morning.

There's no chance of falling back to sleep now, so I turn on the light by the bed and take my tablet out of my attaché case. The last time I updated it was yesterday afternoon. I read it on the plane, but was too tired to update it again when I arrived. I slide the tablet into a docking unit on the desk and display the menu. The hotel offers a choice of several hundred news media from around the world as well as a wide selection of multimedia books and interactive videos.

I could use any of the flat-screen digital televisions in my room to display my selections, but I prefer a tablet for reading. There's something about holding reading matter in my hands and turning pages by touch that feels more comfortable to me. Maybe I'm just old fashioned.

My tablet is an appliance about the size and shape of a standard paper magazine and weighs about a pound. Except for the on/off switch, there are no physical buttons to push. I could attach a keyboard, but mostly I interact with the tablet by touching or writing on the screen. Using a special electronic pen, I can highlight items, attach handwritten notes, and work a crossword puzzle as well as quickly locate items that are stored in my tablet's memory. I can also use a selection of simple voice commands if I choose.

I've preset my tablet always to select *The Miami Herald, CNN HyperNews, The New York Times* and the local newspaper, which in Eugene is the *Register Guard*. My tablet has also been preset to look for stories of special interest to me from among all of the available news media. It will take a few moments to update the tablet, so I go to the kitchen in my room to prepare a cup of coffee. When I return, the dock indicates my selections have been loaded and the battery in my tablet is recharged. I remove the tablet and return to the bed.

When I touch the screen, I am immediately told that 17 stories contain topics that I have included in my personal profile. I'm particularly interested in virtual reality applications, astronomy, and space exploration, as well as news from several South American countries. I could go directly to the stories containing those topics or just browse. Since I have plenty of time this morning, I choose to browse first. I start with the *Miami Herald*'s general news section.

The pages, which resemble traditional newspaper pages, are displayed at about half their normal print size for easy scanning of headlines and graphics. I can "turn" pages without the risk of hitting a person sitting next to me or spilling my coffee. All I do is touch the corner of the screen. Pulling out sections, such as sports and business, is even easier than in print. All sections and departments are listed on every page, so they are always just a touch away.

As soon as I find an item that interests me, I touch it and the complete story or graphic instantly appears on the screen as a readable size. With printed newspapers, readers are always complaining that the text is too small. With a tablet that is never a problem. If I'm having trouble reading the standard text, I can enlarge it to whatever size is comfortable.

If I'm too tired to read, and I need to concentrate on some other activity, such as driving a car or preparing dinner, I can also have the tablet "speak" stories to me. I often use this feature to customize "radio" news reports that I can listen to while I'm driving to the office. With voice commands I can skip to the next story, repeat a story, or stop the report.

A full-color, animated graphic about a hurricane approaching Miami catches my attention. The dateline indicates it was updated this morning at 3:36 my time. I touch the graphic and it enlarges to fill the screen. The three-dimensional animation takes about 10 seconds to show the path of the hurricane with its predicted landfall just south of

Miami. I can replay the graphic again or go directly to the related stories by touching different locations on the screen.

As I'm brousing the *Herald*'s business section, I find a story about the latest federal surtax. An accompanying graphic offers me the opportunity to see how I will be affected. By using the electronic pen to enter my salary and a few other details, the graphic interactively computes and displays the amount of my surtax and shows me how I compare with other taxpayers. I'd like to print the story and graphic later, so I touch "Save" on the screen. This puts the page in my tablet's personal memory.

Another story in the business section makes reference to an event that occurred sometime last week. I apparently missed the story when it was originally published, so I touch an icon for the current abstracts directory associated with this story. This immediately displays the headlines and abstracts for all stories on this subject published by the *Herald* in the past 2 weeks.

After reading this article, I "turn" the page. As might be expected, advertisers are now fully exploiting the advantages of this blended medium. When you touch most ads, they suddenly come alive. Video and sound are often combined with printed messages in entertaining ways. If you request it, some ads will speak to you. Others display their merchandise with short video clips. More importantly, advertisers can deliver a variety of targeted messages that can be matched to each personal profile.

An airline ad offering discount fares to South America attracts me with the haunting music of an Andean flute. I'm planning to take some vacation time in Peru next month, so I touch the ad to get more information. Before I quit, I'll check the ad indexes to see if any other airlines are offering discount fares.

With the built-in communicator, I can even make my reservations directly from the tablet if I choose. The airline's reservation telephone number is imbedded in the ad, and my credit card numbers and other essential data are maintained in the tablet, so all I would have to do is write in the dates and times that I want to travel and touch a button on the screen. The information is encrypted as well as voice-print protected, so there is no risk of someone else placing orders with my tablet.

After browsing the *Herald*, I return to my tablet's main menu and touch the icon for *The New York Times*. The lead story in the "Science Times" section is about the multinational exploration of Mars. By touching an image of the planet, I can watch a 15-second video segment showing views sent several hours earlier by the robot vehicle as it roved the surface of the planet.

Video clips are now a common feature of electronic newspapers.

Many of the still photographs, when touched, become full-motion video with sound, so I can watch and listen to news events as well as read about them. If I missed something, I can play the clip again and even "freeze" an image. Some photographers have worried that full-motion video will make still photography obsolete, but that does not appear to be the case. Both have recognized strengths and are likely to coexist for some time.

I'll be staying over the weekend, so I'll check the *Register Guard*'s "Guide" section to see what's going on. All articles about upcoming events as well as listings and reviews of restaurants, movies, concerts, or books are always available to electronic newspaper subscribers in supplemental guide sections. I can easily select categories of interest and in some cases even see and hear a sampling of events, places and productions. As with airline advertising, many of the ads in these sections offer reservation and ticket purchasing services by way of the tablet's built-in communicator. Some ads even make it possible to order and receive electronic pamphlets and sales catalogues in this manner.

My presentation is at 9 this morning, so I had better take a break from the news and review my speech. Tablets are used for many purposes, not just for news media. In my tablet's personal memory I have stored a copy of my speech and supporting graphics as well as my itinerary and appointment calendar. Using the electronic pen, I can make changes in my speech right to the last minute, which I am always prone to do. At the rostrum, the tablet will serve as my personal prompter.

As soon as my presentation is completed, I can easily provide participants and reporters with electronic or printed copies of the final version. It takes only a few seconds to save an electronic copy on a "memory card." These cards, which are the size and shape of a credit card, have completely replaced the "floppy disks" there were commonly used with early personal computers.

Printing is also easier. To print copies no longer requires cable connections as it did in the last century. Rooms are now pervasively networked, so all I have to do is touch "print" on the screen and indicate the number of copies required. A signal from my tablet will remotely activate the nearest available printer/copier.

Tablets are now so common and essential that it is difficult for me to recall how we managed without them. And when I think about the amount of paper we used to consume and the impact it had on our environment, it seems fortuitous that tablets emerged when they did.

[Editors' note: This example illustrates the full range of capabilities provided by the emerging tablet newspaper.

Typical users might not need all of these features. Many might simply use their flat panel tablet to read the morning newspaper just as they do today and at no greater cost.]

CHALLENGES TO NEWSPAPERS AS A BUSINESS

Despite the disappointments and skepticism, the battle for audiences and market share is being waged with even greater verve and determination in the 1990s. All forms of mass media today see themselves struggling not just against each other, but against a changing economic order, new and potentially formidable competitors, growing social, economic and racial diversity, a failing educational system and declining literacy, and a public that seems increasingly disinterested and disenfranchised. Within the newspaper industry there has been an obvious and dramatic reordering of priorities in response to these real and perceived threats. From its earlier emphasis on content and technology, the industry now appears to be focusing its attention on marketing and alternative sources of revenue.

Throughout most of this century, the media and advertisers had a mutually beneficial and interdependent relationship that seemed to work well and was generally unquestioned. But as the audiences for mass media have become more fragmented and enigmatic, the ability of newspapers, magazines, television, and radio to match advertisers with potential customers has eroded. Advertisers are now questioning the rate structures as well as the presumed benefits of using mass media to deliver their messages. Newspapers and magazines have traditionally based their advertising rates on their circulation numbers. As total circulation increased, so did rates. The rationale used by publishers was that the increased cost of advertising was justified by the increased exposure. To pump up their total circulation numbers, they have tended to use special offers and deep discounts in subscription rates to attract new customers. Who the new customers were made little difference to publishers. However, in the late 1970s, the old pricing formulas started coming under attack. Rather than totals, advertisers began demanding data on the demographic composition of subscribers and pushing for rates based on the ability of publishers to deliver "targeted" audiences. Although newspaper publishers were generally reluctant to provide that data and adjust rates, most eventually created or expanded their market research departments and took steps to attract more of the readers advertisers were seeking.

Historically, newspapers have expanded their audiences by identifying areas of interest to groups of potential readers and then creating new sections or departments to serve them. In the 1980s many newspapers used this approach to create a number of new sections that focused on such topics as local business, personal computing, health and fitness, and entertainment. Although most of these sections succeeded in attracting new readers and advertisers, they also increased the cost of advertising and added to the overall bulk of newspapers. In the same period, however, there was also a general expansion of special "zoned" pages and sections that contained news and advertising focused on individual neighborhoods and communities. Since fewer copies are printed and distributed, newspapers are able to offer advertisers significantly reduced rates and more targeted audiences for these pages, but they still have a number of limitations. In most cases, the lines drawn by newspapers separating one neighborhood from another are arbitrary. Although you might share many of the interests of your neighbor who lives across the street, it is quite possible that your zoned edition of the local newspaper will be different. This problem is accentuated for relatives and friends who live in different parts of town that receive different zoned editions. Even for advertisers, zoned editions risk ignoring potential customers since people often shop in neighborhoods or communities other than the ones in which they live. One of the solutions often discussed by newspaper publishers would be to provide a core newspaper for everyone and then let each subscriber decide which additional special and zoned sections they would like to have delivered each day. The ability of newspaper companies to economically print and distribute daily editions tailored to the needs of individual subscribers presents enormous and perhaps insoluble problems using existing technologies.

Magazine publishers have had somewhat greater success at meeting the demands of advertisers for targeted audiences. Most specialty publications go to great lengths to know who their readers are and what products and services they are interested in buying. Even general circulation magazines have been finding ways to create editions for different regions and different sets of interests.

The struggle to justify their escalating advertising rates has been no less formidable for broadcast television networks and stations. Advertisers are also challenging the validity of standard television rating systems and are seeking greater assurances that their messages are being seen by the audiences they want to reach. With remote controls, cable, and videotape recorders, television audiences are now finding it easier to avoid commercials by "grazing" through programs, which puts additional demands on both the broadcast networks and cable operators to justify their rates.

The efforts by all forms of mass media to meet the demands of advertisers for targeted audiences has had the effect of further fragmenting the media markets. Even without the threat of a new mass medium, the current level of fragmentation is now seen as a formidable obstacle to future growth for all existing mass media companies.

LIVING WITH THE "RELATIVE CONSTANT"

If consumers and advertisers were willing to spend more for mass media each time a new medium was introduced, another slice of the media pie would not be a serious problem. The pie would just get larger. But that is not a choice they seem willing to make. Historically, total spending by Americans for all mass media appears to be relatively fixed, even when an expensive new medium, such as television in the 1950s, has been introduced. Charles E. Scripps, now the chairman of Scripps Howard Newspapers, is believed to have been the first to recognize this constraint on mass media growth. Back in 1959, Scripps referred to his revelation as the "relative constant." Communication theorist Max McCombs has since confirmed and extended Scripps's observations in several studies. McCombs (McCombs & Eyal, 1980; McCombs & Nolan, 1992) has reported that the percentage of U.S. Gross National Product allocated to all mass media has remained relatively constant at 4%-5% since 1929, in good times and in bad.

Not all researchers agree with the conclusions of Scripps and McCombs, but most accept that the fortunes of mass media companies are closely tied to the economy and to the number of media choices available within any time period. In other words, when the economy is growing, new choices are generally less threatening to existing media than when it is stagnant or declining.

If we accept that spending by audiences and advertisers for mass media is unlikely to grow significantly in the future, we must also conclude that any new mass medium that might emerge in the 1990s will have to take revenue away from established media in order to succeed. Since the mass media pie is already thinly sliced, another slice would undoubtedly force media companies to cut their operations and expectations. That may not be possible for many media companies. New technologies are unlikely to further reduce their costs for gathering and processing information leaving merging with other companies or quitting the business as the only alternatives.

There is another possible outcome, one that has been largely unexplored; instead of ever-increasing media choices in the next century, as is generally expected, new technologies and globalization could reverse or stabilize that trend. Where the 20th century has been known for media divergence and fragmentation, the 21st century may be known for media convergence and consolidation. Although ownership of newspapers, magazines, cable, and broadcast radio and television may become more concentrated than it is already, that is not necessarily the most likely outcome of mediamorphosis. With the emergence of each transforming technology, a new wave of entrepreneurs has risen quickly to challenge the slower moving established enterprises; and in many cases they have succeeded in displacing them. The portable digital appliance is a transforming technology, and as such will create opportunities for entrepreneurs to exploit new forms of media.

By consolidation, I am referring more to the processes and forms of content than to ownership. For example, rather than more cable channels offering us more of the same old movies and programs more frequently, media convergence and consolidation can lead to more creative and tailored uses of those channels, ranging from interactive multimedia books to digital music and videos on demand. New media technologies applied by entrepreneurs with fresh visions may well reduce the existing redundancy of non-choices and create opportunities for us to hear more meaningful voices and more diverse points of view.

Recognizing the economics of the "relative constant," it is within this scenario—a greatly expanded diversity of news and information services and the availability of flat panel "digital appliances" offering easy and inexpensive access to these services—that we may best see the needs of our increasingly complex citizen audience served by the newspaper of tomorrow.

Chapter 3

BACKGROUND ISSUES RELATED TO INFORMATION SERVICES

BY MARK A. THALHIMER

As the concept of a national information service was pursued as a Center research project, it was important to consider many of the background issues associated with electronic information services. These issues were summarized in various reports accumulated by the Center staff from which this present chapter was compiled. Among these issues are contemporary developments in switched network telecommunications, existing electronic information services, video "dial-tone," information services as a public resource or utility, regulatory implications, types of initiatives for establishing a national service, and commerical implications. This chapter in an earlier form was one of the "background" papers for the Roundtable reported in chapter 4.
Mark A. Thalhimer is technology manager at The Freedom Forum Media Studies Center.

> Whether information media enhance or retard freedom and culture depends on the rules under which they are allowed to operate.
>
> —*de Sola Pool* (1990, p. vii)

IN THE PUBLIC INTEREST?

Since 1984 and the divestiture of the American Telephone and Telegraph (AT&T) Company, there has been a raging battle mainly between newspaper publishers and telephone companies as to who should deliver electronic information services to the public. The post-divestiture AT&T was allowed to enter the business after a 7-year waiting period, whereas the divested Bell Companies, the operators of locally franchised telephone services, continued to be barred by provisions of the antitrust decree. In July 1991, the legal tide appeared to be moving in favor of local telephone service providers when Federal Judge Harold Greene lifted restrictions on the telephone companies ability to provide information services. Although he immediately placed a stay on this decision until all appeals could be heard, in October 1991 the Appeals Court reviewing Judge Greene's decision removed the temporary stay, thereby lifting the judicial barriers that had kept the telephone companies from originating and owning the information it wished to sell. The ownership issue is important because the local telephone companies were not barred, at least nationally, from selling "conduit" services to other information providers. Although there are wide-ranging arguments—some reviewed in other chapters of this book—about whether the company that owns the conduit should also own the content, a truly practical fear of the newspaper publishers was that telephone companies would move aggressively into the advertising market as an expansion of *Yellow Pages* ™ and thus cut into one of the remaining positive revenue streams of newspapers, namely, classified advertising.

In the time up to the writing of this chapter (1993), the situation still remains cloudy. Bills have been introduced in Congress to re-regulate certain aspects of telecommunications, including information services. Even some of the states have gotten into the act in attempting state-level regulation that would affect information service by local telephone companies that are franchised for operation under the dictates of state public utility commissions. Regional Bell Operating Companies that own the local Bell telephone companies have not only lobbied agressively to bar such regulation but have moved strongly in the direction of developing unregulated branches of their companies that supposedly could offer information services without any utility commission involvement.

Although all parties involved in this dispute have constantly appealed to "public interest" arguments, neither the arguments they have voiced nor even most of the services they have demonstrated have truly been exactly in the public interest. Debate has been more the product of politicicans, attorneys and lobbists than from individuals who understand the information and transaction needs of the public at large or who can envisage the technological opportunities of what could literally be a new public medium. The situation has caused one to recall similar lobbying efforts in the days of film versus television, broadcasters versus cable, the film industry versus cassette distributors, or broadcasters versus direct broadcast satellite operators. We have seen mainly well-financed corporations and their industry trade groups represent special economic interests as they relate to provision and sale of information. The judicial, legislative, and regulatory areas of government have been besieged by the well-heeled lobbyists and attorneys fighting over access to the information services market. The needs or desires of the consumer, or more abstractly the information need of the citizens of our democracy, have been largely overlooked. All of this has raised the question: Whose public interest has been addressed, the information providers' or the citizens'?

Indeed, what if one did turn to look to the interests of the citizen with the perspective of electronic information services being regarded as a potentially new public medium? The very prospect of this questions raises many issues, from the importance of telecommunications advances, to interpretations of the First Amendment. The issues are worthy of at least some modest consideration if we are to consider a true public interest perspective. This chapter reviews briefly a number of these issues.

THE SWITCHED NETWORK AS A "PLATFORM"

When most Americans hear the term *telecommunications* they think of getting messages electronically from one point to another, such as with a telephone call, a telegram, a "fax," or even as the method by which they receive radio or television broadcasts either over the air or via cable. At a slightly more detailed level, there are many distinctions among these services, although they are all broadly telecommunications. Important to our present discussion is the concept of a *switched network*, which is the key technology underlying a national information system. Basically, the largest example of a switched network is the public telephone network that links all of our telephones via central offices and long distance connections to each other. A switched network allows us to connect and disconnect telecommunications users to one another simply at the press of a few buttons. By contrast, telecommunications which serves television and radio broadcasting is largely a one-way, fixed communications system. We cannot communicate by radio or TV image back to the broadcasting station or to one another, given the present broadcast or cable television technologies. By contrast, the switched network is important to our present discussion because it allows us to connect our telephone devices not only to one another but to computer-based services that now can provide us with text, graphics, even simple visual images and eventually full video services.

Most of the American public also does not associate telephone service with other electronic media, especially radio and television, in part because of legal decisions that have divided up the telecommunications services industries in the United States. A main example is the American Telephone & Telegraph Company's agreement to exit the radio broadcasing business in the 1930s. This left broadcasting development to other major companies while switched network development was the main business of AT&T in this country. In 1982 a further legal decision divided America's switched network business essentially between long distance services, which AT&T would continue to offer, and local telephone services which would be the exclusive monopoly domain of local "Bell" and other independent telephone companies. This decision, called the Modified Final Judgement, established seven Regional Bell Holding Companies which, in turn, operate local Bell telephone companies. The remaining downsized AT&T was allowed to provide long-distance service, and to enter into manufacturing and computer-related businesses.

The Bells were given the regulated monopoly on local telephone service in their respective areas but were prohibited from three areas of business; namely manufacturing, long-distance telephone service, and providing information services beyond directory information and the printed Yellow and White Pages directories. The holding companies are:

NYNEX – New York and New England;
Bell Atlantic – New Jersey to Virginia;
BellSouth – North Carolina to Arkansas and Louisiana;
Ameritech – Upper Midwest;
Southwestern Bell – Texas and the lower Midwest;
Pacific Telesis – California and Nevada; and,
U.S. West – Northwest, some Rocky Mountain states, and across to Minnesota and Iowa

These companies, or more specifically their subsidiaries, already offer some information services. Directory assistance, for example, has been available almost since the beginning of telephone service itself. Depending on the companies, there has traditionally been such services as time-of-day, the weather, and even theater information. (Years ago in Hungary, you could dial up an "A" pitch to tune your violin.) In the last several decades, with the advent of tone-code dialing, there has been the growth of many "dial menu" services such as health counseling, guides to restaurants, transportation schedules, and the like. Guides to these can now be seen in most local telephone books. Often referred to as "audiotext," these services have been growing, but not mainly as a major new medium.

Also there is a whole near-term history of what most call "videotex," where various information, entertainment, and message exchange services have been accessible via special terminals connected to the telephone network. Newspaper ventures into this area, such as Knight-Ridder's "Viewtron," or Times-Mirror's "Gateway," were financially unsuccessful. Companies like Prodigy and Compuserve offer public access to such services but their total customer bases are not a mass market; mainly they are in the low millions in subscribers. This makes the information providers more akin to a "magazine" subscription provider in customers than a national information service.

There are also, as described by William Dutton in chapter 5, a wide variety of local information services that use the telephone as the access technology. Although researchers like Dutton see them as a growth area, it would take national policymaking to make them a major communications medium.

So why the issue of a national information service? Is there the possibility of anything new in this area, or just more of the same? One major milestone or "opening" that we have sensed in our research into this topic is that several of the major bottlenecks to the development of information services as a new medium are now being eased. One is the capacity of the telecommunications network. With the widespread deployment of digital switching, it is not only much easier but much more cost-effective to offer services that require computer database storage, retrieval, and transmission. Digitally managed networks operate in the same "language" as computer codes so the combination of network and computational technology is greatly facilitated. Another reduced bottleneck is compression technologies that allow much more information to be transmitted and at a faster rate than on a traditional telephone line. Some compression technologies allow much more efficent use of existing "twisted pair" copper telephone lines originally developed just to carry as voice call but which now with advanced technology can carry a reduced-quality video image. Other applications are already developed for higher capacity telecommunication media such as fiber optics which, for example, could carry multiple telephone calls, video services, and a wide variety of information and other electronic services via one "conduit" into the home.

There is also the possible demise of the bottleneck of having to use a computer, or "dumb" terminal, to access text and graphics over telephone lines. This advance is represented in one example by the coming of the so-called "smart" phone. New telephones are about to be marketed that come equipped with a significant amount of computer intelligence masked by an easy-to-use set of control buttons, a small visual readout and a fairly high-quality touch-sensitive viewing screen. The display might be similar to what is used in bank automated teller machines. This new phone allows standard voice communications along with expanded key-push messaging which gives it easy-to-user controls for sending and receiving information, requesting services, or carrying out electronic transactions (e.g., writing checks). Smart phones are being designed, and eventually priced, to give customers easy access to new electronic services without the need of a person feeling that he or she has "to use a computer" to gain an information service.

Under development are more advanced telephones or terminals, or software for existing high-end home computers, that allow them to operate as multimedia devices accessing voice, text, and even television programming. Those who can or are willing to invest in a more sophisticated "telecomputers" would have enhanced access to network services existing above the "guaranteed minimum." The market will develop the higher end, expensive, computerized communications devices, like a wide-screen high-definition viewing system, and those who can afford to use them will do so.

Eventually, some form of switched "video dial-tone" will likely be available through the telecommunications network. This could add to the multimedia capabilities of a national service. Video dial-tone would mean that telephone users could make a "video phone call" between two places as easily as they can make an audio telephone call today. The objective would be to allow video transmission with the same ease that we currently have audio transmission. Consumers would also be able to connect to video databases or "hosts" and retrieve video in the same way that computer users connect to on-line databases today, to retrieve text information.

Video dial-tone has been on the federal policy agenda. Former Federal Communications Commission Chairman Alfred Sikes spoke frequently of the importance of developing video dial-tone services. In a presentation (Sikes, 1991; see also chapter 7) early in his administration, Sikes applauded the concept of "video dial-tone" as a way to provide "a window on the world" for the average telephone customer. Sikes said video dial-tone should be "a national goal" to be provided "as early as customer demand warrants." Although markets will spur development of such technology, Sikes added, "the federal and state governments must take the necessary steps to encourage the investment in research, development, manufacturing and construction that will be necessary." While reiterating phrases about the importance of market competition, Sikes sounded a note of warning that the struggle among competing providers is not always good for the development of communications. "The essence of this conflict surrounds distribution and especially who is going to control tomorrow's 'information highways' into the home. The risk is that this battle is going to blunt progress," he said. Sikes' vision is of a world where vendors "should be able to offer subscribers easy video access—a means of getting various video services when and how they choose." Sikes said he could not "think of any development that would place more power, more freedom, in people's hands than switching on a 'video dial-tone.' The transformation would represent a quantum leap in individual choice and capabilities." Access to video over the public switched network will probably be some years in coming to full fruition but there seems little doubt that it will come, and with it eventually are prospects for a full range of multimedia access over the network, including three-dimensional and "virtual reality" interactions.

The policy "message" from Washington, along with trends toward local deregulation of the telephone business, are yet another example of a disappearing bottleneck. Whereas most state-level regulation has focused only on keeping local telephone rates low, there is a growing interest in telecommunications as an economic development tool for creating new businesses and jobs (e.g., see Schmandt, Williams, Wilson, & Strover, 1991; Williams, 1992). Information services, including both commercial and public service applications at the state and local level are one of these developmental areas. Such growth areas are topics of state studies and planning that go beyond the typical interest of state utility commissions.

But despite these grand images from the future, we are mainly concerned in this chapter with the prospects of network information services for the present or very near future. Unless, there is national policy to steer network services development as a broad publicly accessible medium, it is likely to develop the foregoing futures for only the priviledged few in our society.

INFORMATION SERVICES AS A PUBLIC RESOURCE

In the sense of serving the public's need for information of all types, including news, an electronic information service can be thought of operating, at least in part, as a member of the press. If so, should it be guarded by the same rights the Constitution and courts ascribe to our traditional newspapers? Unfortunately, our country has no guiding vision of how different communication media should interact with each other and with society. On this topic, Everette E. Dennis, executive director of The Freedom Forum Media Studies Center, wrote, "there is a natural tendency in the United States to eschew a communications policy, letting our 'policy' be whatever the fragmented and often chaotic yield of executive order, legislation and rule-making might make it. ... now there seems to be a greater need to clarify the guidelines that protect citizen interests" (Dennis, 1988).

Services received through the switched network will likely eventually occupy a place alongside print and broadcast as a medium of news communication. Should these new network-based services be given the First Amendment protections of free speech and press? Lawrence K. Grossman (1991), former president of NBC News and the Public Broadcasting Service, wrote:

> Democracy will best be served in the 21st century by returning to the 18th-century idea of an independent and totally unregulated press, a press that is controlled by many different owners, a press that offers access to many different voices, and a press that makes available essential public affairs, educational and cultural programming to all our citizens.(p. 72)

If we are considering service as a public good, should a national information service be regulated as a *public utility*? A national interactive multimedia network will reach its full potential only if all citizens are given access to the communications and information services it provides. The system must be monitored and controlled by the widest possible range of citizens.

Elected officials, the courts, and society have often had difficulty applying existing legal principles to new technologies. One basic legal principle is the First Amendment to the Constitution, which states, "Congress shall make no law . . . abridging the freedom of speech or of the press; or the right of the people to assemble, and to petition the Government for a redress of grievances." Access to greatly expanded network information services will have such a far-reaching impact on our society that the network may eventually become the primary means for citizens' free speech; access to the press; means of electronic assembly; and method of petitioning the government, to reflect key First Amendment terms. As such, there are likely regulatory implications to ensure the public good of a national information system, our next main discussion issue.

REGULATORY IMPLICATIONS

In addition to government regulation derived from the First Amendment, electronic means of communication have already been regulated to administer scarce resources and to address certain public needs. There is only a finite amount of broadcast spectrum available in any geographic region for radio and television. Today, the broadcast media must compete with other uses of the spectrum, like cellular telephone and paging services.

Current federal regulation of the electronic media grew out of a series of statutes that were often created as a protective barrier to restrict the media rather than as an incentive to create new methods of communication. A good starting point to describe communications regulation is the Interstate Commerce Act of 1887, which established the Interstate Commerce Commission (ICC). Its primary mission at the outset was railroad regulation. The Mann-Elkins Act of 1910 placed the common carriers, both the telephone and the telegraph, under the auspices of the ICC, but the Interstate Commerce Commission did not actively regulate the new communications systems. During the 24 years of ICC supervision over these systems, the Commission heard only 14 cases involving the telephone and telegraph.

The first appeal for regulatory restrictions on the broadcast airwaves came from the United States military. The U.S. Navy had used radio as a tool for ship-to-shore communications in the early 1900s and argued that it should have a monopoly on radio communications as a means of controlling the broadcast spectrum. One incident, in particular, prompted the Navy's concern about civilian use of radio communications. In 1906 a radio station operated by the General Electric Company in Brant Rock, Massachusetts broadcast Christmas greetings to the U.S. Navy radio technicians on ships at sea (Emord, 1991). Congress did not give the Navy a monopoly on radio communications, but the Radio Act of 1912 restricted the use of radio transmissions to those with a license from the Secretary of Commerce. The Act also reserved some parts of the spectrum for governmental use and set guidelines for the use of the "airwaves" and emergency transmissions. This system worked well to regulate radio usage until the early 1920s. Then in November 1920, the first commercial radio station, KDKA, began operating in Pittsburgh, where it broadcast the first mass media content for a wide consumer audience. Within only a few years, hundreds of commercial radio stations were on the air.

In 1923 a federal court ruled that the Secretary of Commerce could not withhold a license, even if issuing a new one would cause broadcast interference with previously existing radio broadcasting stations. After several years of broadcasting chaos, Congress passed the Radio Act of 1927, which established the Federal Radio Commission (FRC) and allowed it to regulate the licensing and re-licensing of applicants based on transmitted content and their service in the, "public interest, convenience and necessity." The Commission had the power to revoke licences but rarely did so unless broadcasters' content failed to meet with the Commission's standards for public service.

Congress enacted the Communications Act of 1934 to expand the powers it had given to the FRC. The Act created the Federal Communications Commission (FCC) to replace the FRC and broadened the new Commission's regulatory powers to include telephone and telegraph communications. Through the process of license application and renewal, the FCC maintains loose control over what may be broadcast over the airwaves. The intent is to maintain a balance between freedom of communication and competing public interests. To do so, the FCC has developed five discrete criteria to judge an application for renewal. Those criteria are (Emord, 1991):

1. The licensee's efforts to ascertain the needs, problems and interests of its community.
2. The licensee's programmatic response to those needs.
3. The licensee's reputation in the community for serving the needs, problems and interests of the community.
4. The licensee's record of compliance with the Communications Act and FCC rules and policies.
5. The presence or absence of any special effort at community outreach or toward providing a forum for local self-expression.

Should such criteria be applied to new policy regarding electronic information services being provided for the public good? The goal of these criteria was conceived broadly to encourage those who hold broadcast licenses to serve the public in their transmission area. Debate has focused on the Commission's effectiveness: Do these regulations serve public needs or merely maintain a *status quo* of broadcast license holders? FCC regulation of the telephone and telegraph industries has not generally dealt with content. Due to the nature of point-to-point communications, the common carrier acts as a conduit connecting any two or more users. The telephone company is not concerned with what is communicated over its network. The common carrier will monitor content only under a court order or when obscenity or harassment is involved. This may change as the telephone companies begin to deliver information services to customers.

A key issue in policymaking concerning a public information service is its status as either a common carrier or a content-provider.

CABLE AND OTHER MEDIA RESOURCES

The cable television industry exists at a regulatory and technological nexus, halfway between the broadcast industry and the switched telephone networks. More than 80 % of American households have the option of receiving cable television. Today, 62 % of the 92.1 million existing American households do subscribe to cable services (57.1 million households) and 77 % of American households have a remote control (70.9 million households).

The capacity to send as well as receive will be important if cable is to offer information services. Those services might include customer-directed video on demand, electronic messaging and information retrieval, and automated gas and electrical meter reading services through the television or a connected device. There may even be a time when the cable companies will offer voice telephone service over the cable systems, in competition with the local telephone companies. Of course, the telephone companies may want to offer the same video and programming services that cable companies offer.

Most cable systems are regulated as franchises within a specific area. Cable franchises are awarded on the state and local level, and franchisers receive an exclusive right to lay cable in a specific geographic area and sell cable television services to the residents of that area. Usually, the decision as to which cable company will receive a franchise hinges on the public access/public service aspect of the proposal. The franchise board justifies the awarding of a local monopoly to a particular cable company based on the company's proposed plans for building up the cable network infrastructure, its ability to pay the tariffs and fees that will be assessed upon it, and the level of public access/public service to be provided to the community.

The most recent, significant piece of legislation affecting the cable television industry is the Cable Communications Policy Act of 1984. It states that all cable systems must have a franchise to operate. The Act also put state and local authorities firmly in charge of the franchising and renewal process for cable systems. Cable rates were effectively deregulated with only basic cable rates remaining under some regulation. The cable operators, in fighting for rate deregulation, had argued that other television sources (broadcast, VCR, and other technologies) provided effective competition to cable services. Others, however, argued that cable systems are "effective" monopolies and should be regulated. The result of deregulation is that cable rates have increased dramatically over the past several years. A report released by the congressional General Accounting Office in July of 1991 said that cable prices had risen on average 56 % since the industry was deregulated in 1986 under the 1984 Cable Act.

The tradeoff for the lifting of rate regulation was access to channels. Under the Cable Act of 1984, cable operators must provide channel capacity for public, education and government use. These public-access channels are now part of any franchise proposal or renewal request. The number of access channels often depends on the total number of channels in a cable system. For example, a cable system with 30 channels might be required to provide 2 or 3 of them for governmental, educational, and nonprofit uses while a cable system with 100 channels might have to set aside 10 for use by nonprofit, governmental, education, universities, and public-access institutions. The cable operator may also be required to build or provide studio facilities for public-access use. Unfortunately, in the scramble to win franchises, cable operators often promise "community service" access that is far beyond what is economically feasible or practical for them to provide.

Many new communication technologies are entering the marketplace at this time and could be relevant to a national information service. Direct Broadcast Satellite (DBS) has been proposed as a means of transmitting television directly to the consumer via satellites. DBS has not had much success in the United States, in part due to the wide range of choices already available to the consumer. The technology may have more success in countries with a less saturated media marketplace. Multipoint Distribution Service (MDS) and Multichannel Multipoint Distribution Service (MMDS) are technological systems that distribute programming via direct line-of-sight microwave signals. The signal is received by an antenna and sent by cable to the user. This system is often used for buildings or communities that do not have an existing cable system and sometimes as competition with existing cable systems.

ORGANIZING A NATIONAL SERVICE

The creation of a national information service can provide a reliable and easy-to-use source of information serving a diverse set of citizen needs while at the same time maintaining an open, free market system that will foster innovation and creativity. However, a number of complex issues surround the creation of a such a service. Although this chapter is primarily concerned with the content that will be available through the switched networks, issues of ownership, operation and available technologies are important because they influence the overall development of a national information service.

Ownership and Operation

Which companies or organizations should be encouraged to operate the new switched network communications services? Currently, organizations competing in the information marketplace include the telephone companies, cable television companies, broadcasters, television and motion picture studios, and newspaper and magazine publishers. It should be noted, that with the exception of the telephone companies, there is a great deal of cross ownership among the other media companies (e.g., Time Warner Inc. has operations in cable television, programming development, videocassette distribution, new interactive media, and well-known print publications such as *TIME* and *Sports Illustrated*).

New Technologies

In addition to the complexity of ownership, media corporations' interests are often stifled by a tangle of competing technologies. Media enterprises must constantly weigh competing technologies as a vehicle for release of programming or "software." Should a new movie be released in the movie theater first or should it go directly to a cable premium channel? Should it go to a cable pay-per-view channel or should it be released on videocassette first? New technologies (including the switched network) will only complicate the variety of choices available to media executives. For the public, however, these new choices have the potential to offer consumers a wide variety of options never before available.

One example of this technological debate is high definition television (HDTV), which is slowly evolving through the Federal Communications Commission's standards testing and evaluation procedures. The FCC's intent, over the long run, is to set a new, high-quality television transmission standard. No one can predict how the new standard will affect existing broadcasters and their spectrum allocations, the cable television industry, or emerging services available over switched networks. Finally, will the American consumer adopt HDTV if and when it does enter the marketplace?

Copper Versus Fiber

One primary technical debate in networked communications concerns the economics of laying copper telephone cables versus fiber optic cables. Relative to capacity and multimedia capability, this debate holds implications for a national information service. Many of the nation's long-distance telephone lines that were originally copper have been replaced by fiber optic cable. At the local level there are millions of miles of copper wires beneath the streets and on telephone poles. Should copper wire in the local loop be replaced with fiber optic cable? A report released this year estimates the cost of an accelerated deployment of fiber optic cable to the home at $200 billion over a period of 20 years. Who will pay for this re-cabling of America? Who will be allowed to use the conduit? Will the results to the citizen be worth the price paid through rate hikes?

Events currently underway in Japan may provide a useful example. The Nippon Telegraph and Telephone Corporation (NTT), which has a monopoly on telephone service in Japan, expects to spend $60 billion to connect all 45 million Japanese households with fiber optics by the year 2015. In an interview with the *New York Times*, Masashi Kojima (Ramirez, 1991), president of NTT stated, "Perhaps this is inappropriate for the head of a telecommunications company to say, but... I wonder if this flood of information really makes our life richer (p. 18)".

As an alternative to laying fiber optic cable, researchers are developing compression techniques to send more information through the narrow transmission capacity of existing copper wires. These methods of compression may make it possible to transmit a video signal over the telephone companies' existing switched networks.

EXAMPLES OF NATIONAL INITIATIVES

The development of our national highway system and public broadcasting system may be useful examples for consideration in the development of a national information service (NIS). In the early 1950s the Eisenhower administration began planning for a national highway system. The Federal-Aid Highway Act of 1956 authorized construction of the express highway system and "hearty and regular increases in aid for building urban, primary, and farm-market roads." The federal government paid for most of the highway construction by giving the money to the state governments. The highway construction was partially self-financing through taxes on trucks, buses, and motorists. The distribution of federal seed money at a state and local level may be a useful way of promoting new telecommunications-based services.

A second example for the NIS is the Public Broadcasting Act of 1967. President Johnson founded the Corporation for Public Broadcasting with the goal of making educational radio and television service available to all the citizens of the United States. Public broadcasting remained on shaky ground for several years due to financial difficulties. Finally, the Public Broadcasting Financing Act of 1974 was enacted to develop a permanent source of funds for public broadcasting. The public broadcasting system is a loose affiliation of television stations that are licensed to nonprofit community organizations, state education departments, universities, and other groups. Public broadcasting at the national and local level is funded through a complex mix of public and private sources.

COMBINED COMMERCIAL AND NOT-FOR-PROFIT?

Commercial services could be published through electronic Yellow Pages developed by the telephone companies or other private providers. Perhaps a low tax on advertising listings in the electronic Yellow Pages could be used to pay for the noncommercial NIS listings. In the NIS, most of the information the public would need could be broken down into a few broad categories, such as health, social services, economic, community, and government. The categories would branch out into various subcategories of information. The categories might initially be accessed through audiotex response systems and later through a simple keypad attached to the telephone system. This basic system could be easily expanded to accommodate new information as needed. Following are general lists of information and how they may be organized:

Community
community services, agencies, referrals
environmental hazards (weather reports,
poisons, and treatment)
public transportation schedules
telephone directory

Government and Education
municipal services, announcements, and meetings
library services
real estate, land records, and deed registration
school district news, announcements, children's
grades and records

Economic
banking and paying bills
credit services and rights
divorce laws, rights and referrals
financial information
immigration laws and rights
insurance terms and tips
job listings
legal terms and tips
utilities information

Health, Emergency and Social Services
abortion rights and counseling
alcohol abuse
child abuse
drug abuse
hospital guide
nutrition
medical emergency tips, referrals, and appointments
nutrition
pregnancy counseling and tips
rape counseling, tips, and referrals
spouse abuse

News Headlines and Stories
classified advertising
more extensive weather reports, international news
(all categories)

These services could be cross-indexed so that a user looking for current information on a health topic like child care would be guided to related governmental and community services. The possible combinations of information and needs are endless.

A national information service should serve the broad goal of allowing citizens equal access to the information they require for daily life. As commercial services are developed for the switched network, a portion of the new medium could be reserved for public use and citizen services. If all citizens do not have equal, effective access to a *minimum* level of services on the network, our society will continue to be divided along economic lines.

A primary goal of this project is to stimulate debate over what services might be made available over the switched network and how all members of the U.S. population should have access to those services. Our national highway system and public broadcasting services were developed from limited resources to serve a national goal. A small allocation of resources to develop national information services should go a long way toward leveling current inequities that limit many citizens' access to important information. While the free market is a good mechanism for developing new technologies, it should not be the only way through which we allocate news, information and knowledge in our society.

DEMOCRATIZATION

If current deregulatory trends continue, network-based information services are likely to evolve primarily along the supply-and-demand laws of the marketplace. Currently, the benefits of information networks are confined to commercial uses such as stock quotes, news reports, electronic mail, and "bulletin boards." Access based solely on the ability to pay would continue to keep many citizens off the network and outside the daily exchange of ideas and information. Powerful telecommunications technologies would remain a tool for the elite segments of society that can afford them.

A national information service could democratize information by making it accessible to a wide range of citizens. Opening the information gates would give more members of society an equal voice. Voters could use the NIS to communicate with public officials; children could participate in school lectures; and citizens could receive the latest news and weather reports. The network could provide local government and school announcements, as well as advice on legal, medical, or financial matters.

The conflict between public needs and commercial viability should not be allowed to reduce broadbased citizen access to information.

Chapter 4

ROUNDTABLE: SIZING UP PROSPECTS FOR A NATIONAL INFORMATION SERVICE

BY JOHN V. PAVLIK AND MARK A. THALHIMER

This chapter summarizes the highlights of a lively debate on whether the concept of a national information service is a worthy topic for continued study, and if so, what are the important issues and major themes? The parameters of the present debate were first outlined in a policy paper: "Priorities for a National Information Service in the United States." That work subsequently led to a research roundtable at the Center in early 1992 on "Citizen Information Needs and a National Information Service," bringing together a diverse group of 22 scholars, industry representatives and policy analysts in telecommunications and media. The chapter includes comments from Julius Barnathan, Leo Bogart, Anne W. Branscomb, Stuart Brotman, John Carey, Benjamin Compaine, Jannette Dates, Everette E. Dennis, Roger Fidler, Patrick Garry, Lawrence Grossman, K. Kendall Guthrie, Susan Hadden, George Heilmeier, James Hoge, Donna Lampert, John V. Pavlik, Fred Tuccillo, Jorge Schement, Alice Tait, Mark A. Thalhimer, and Frederick Williams. Also included is an addendum from Bellcore president, George Heilmeier, describing the Advanced Intelligent Network, which could likely affect the development of a national program of citizen information services. Please see Appendix for the participants' affiliations.

OVERVIEW OF THE ROUNDTABLE

Few involved in communication and media can escape the escalating debate over emerging electronic information services and questions about the structure of the brave new information world embodied in the new public communications medium. Pressing questions arise over where a new information service will leave existing communications media—print, broadcast, cable and telephone—and over what their roles will be and who will control what parts of the vast new electronic market.

What's missing in these provider-based debates, however, is thoughtful discussion of how the emerging telecommunications medium might benefit society and individuals. Given trends toward regulatory flexibility in telecommunications, there is a real danger that neglecting consumer and constitutional issues could result in a system of competitive services at the exclusion of many public-interest priorities.

One goal of the roundtable was to gain a "high ground" perspective on the implications an interactive electronic information service might have for an increasingly complex and multicultural society. Such an information network, made possible by convergence of already available telecommunications and computer technologies, would provide citizens with instant interactive access to a staggering array of information databases and services. Questions of content remain: What kinds of information services might benefit all citizens? How could such a system be structured, implemented, and funded so as to ensure universal access and retard or reverse ominous trends toward a separate and unequal society of information haves and have-nots? And how could the system provide optimum public service elements while remaining financially attractive to operators and information providers in a competitive marketplace?

Telecommunications and media analysts predict that the convergence of computer and telecommunications technology will result in creation of what effectively will be a new communication medium. It will be a relatively short time before the technology exists to package a variety of communications media, including audio, graphics and text, into a single fiber-optic telephone line. Connected to the telephone system, large-scale computer networks will be able to receive, store, repackage and distribute information customized by consumers to their individual needs, transmitted on demand simultaneously, interactively and globally.

Information vendors—including telephone companies, cable operators, broadcasters, newspaper publishers, and others—will provide a range of services for the multimedia network. Consumers will be able to satisfy their information needs individually, selecting from vast menus of choices. Mechanisms that determine survival in a market economy will work their way, and information services deemed of declining value will be replaced by better services, offering the opportunity of widespread public access to a global network of news, resource, and entertainment data.

Although the technology is not yet fully in place, industry players are already jockeying for position in a race for control over what they believe to be an inevitable and profitable medium of immense promise and reach. Individuals are all but lost in these maneuverings, represented only by faint voices with disparate goals at best. There is the not-unfounded concern that, given free reign, commercial vendors would seize control of this medium before any public policy might be developed to reserve a portion of it for public use and citizen access. Consideration of the capabilities and promise of a national and international public-access information network is an opportunity to evaluate the implications of the effect of this emerging technology from the point of view of individual users and larger societal needs.

Information has long been a consumable product, packaged and marketed for a price. Arguably, this characteristic, facilitated by sophisticated technology such as personal computers and cellular phones, has increased the privatization of information to the point where the phrase "free speech" may become a threatened concept. Many assert, however, that certain kinds of information must remain available to all—universal access to 911 emergency service and other hotlines, for instance. These lines of communication are funded by methods other than per-use charges, acknowledging that individual ability to pay is secondary to the overriding public interest of providing for citizens' safety and health.

Clearly, these new technologies portend dramatic economic changes for those with financial interests in the media marketplace, hence industry focus on market control. But complex questions remain about the sociological implications of the new communication medium and the mechanisms by which all citizens will be guaranteed full First Amendment rights. Further, concern exists over how to assure equal access to the basic information services available through this technology, regardless of socioeconomic status.

Take, for example, the case of a "smart" phone, which, for less than $200, could provide users with switched interactive transmission of voice, data and video. But even this relatively modest sum may well be beyond the means of many in society, especially those who may be in the greatest need of information. For these, policies could be developed making public smart phones available, permitting access to public information through use of a universally issued "smart card." Similar to an encoded bank card, this would contain information—the user's social security number or driver's license number, for example—which would secure access to a public information terminal.

The Freedom Forum Center for Media Studies has undertaken a long-term project to explore the social, economic, and political implications of these emerging information technologies. The research roundtable, convened in early 1992, was a first step. The roundtable's premise was that any citizens' information service should have as its primary goal equal access to the information citizens require for daily life. Panelists approached the subject by developing the discussion around such issues as: defining the information marketplace;

asking who needs information; localization of services; challenges of access; how policy might be formed: insuring fairness, privacy and responsibility; existence of policy models; costs and funding questions; and cooperation among stakeholders.

DEFINING THE CITIZENS' ELECTRONIC INFORMATION MARKETPLACE

The roundtable discussion began with an inventory of basic information needs that might be met through a public-access information system. From broad categories of information—"community information," "government and education" or "emergency, health and social services," for instance—users could navigate quickly into subcategories for answers to their specific information needs. These might include such diverse information categories as "public transportation schedules," "school district news," "immigration laws," "job listings," "health-care providers," "rape counseling," "real estate records," "abortion services," and other arguably essential information.

Users looking for information on child care, for example, would be guided through cross-indexes to related government and community services, said Mark Thalhimer of The Freedom Forum Media Studies Center. "The possible combinations of information and needs are endless."

But limited resources, including technology, money, and channel access, are likely to restrict these endless possibilities. Assuming that any national information service should retain a public interest component, policymakers must define at some point precisely what kinds of information citizens need. The idea of dictating public needs sparked controversy, with some panelists rejecting paternalistic approaches toward defining what people need in favor of examinations of what kinds of information people want. "This reminds me of the old public TV debate—the good liberal elitists, sitting around the table telling everyone what they need," said Benjamin Compaine, president of Samara Associates of Cambridge, Massachusetts.

"When people do have specific needs, they pretty much find a way of getting them, whether it's by word of mouth or tracking them down," he added. "The telephone works wonderfully for these things. I say we should be focusing on wants rather than needs."

Historically, much of the debate around development of U.S. communication media policy has been a series of efforts by government to resolve the conflict between consumer demand and what is perceived as necessary for an informed public. Information, as a commodity, may be defined in terms of market demand, but there may well be other kinds of information, less commercially attractive as commodities, that are considered by the larger society as essential to help individuals function as parts of that society.

K. Kendall Guthrie, a policy analyst for the Department of Telecommunications and Energy of the City of New York, pointed out the government responsibility in this matter. "Local governments have a huge amount of information that people may not necessarily want, but they definitely have to get at," she said. "Like what kinds of information you need to bring to get your driver's license when it's stolen. What the mass transit schedules are. How to get a marriage license."

While perhaps not hot market commodities, these kinds of information are the responsibility of government to maintain and keep available to the public. This kind of public record data—procedural information for dealing with government agencies, as well as listings of services and facilities—are reasonably stable and relatively inexpensive to maintain. Other valuable resource information, however—such as telephone numbers, public notices, and alerts and other time-bound data—change constantly and require perpetual updating. This is "information that makes society run better but needs a better distribution system," Guthrie said. "This would be information like traffic alerts, so that when people left their offices, they could get some information on where the traffic jams are and how to spread the traffic out."

Providing this information predictably and reliably could enhance more than a few aspects of maneuvering through modern society.

WHO NEEDS INFORMATION, AND IN WHAT FORM?

Although some individuals may require only intermittent access to resource data, an equitable information service should consider the needs of all potential users.

"I think when we say that some of these information needs are sporadic, we're not thinking about the poor," said Jannette Dates, a Center fellow and associate dean of the School of Communications at Howard University. "Things like job training, skills training, availability of jobs are constant concerns for people who are poor or unemployed."

Attempts to define user needs may well yield a core of unanimously sanctioned information. Beyond that core, however, lies much room for debate, where information that one individual, group or institution may see as essential is superfluous to others. As some panelists pointed out, not everyone knows all his or her information needs, or has a very firm understanding of what kinds of information is needed, or when.

Newspapers have tried to define this question of information demand in terms of "news you can use," as Fred Tuccillo, assistant to the editor of *New York Newsday*, said. Readers don't always know what news they can use, he said. "The problem with newspaper-delivered information is that the time the person needs the information may not coincide with the day it's published," he said.

The question is clouded further by the distinction between information that people actively seek and information they find is useful almost by accident, said Leo Bogart, an adjunct professor of marketing at New York University and former longtime executive director of the Newspaper Advertising Bureau. "A large part of what we find useful is in fact information that we would never seek out on our own from an available database," Bogart said. "It's when *USA Today* puts that little graph at the bottom of page one and we say, 'Ah, I didn't know that.' That's something that we can use."

Policymakers attempting to determine what is necessary information in a given community might begin by examining the information needs already met by existing sources, suggested Frederick Williams, a professor of communications at the University of Texas at Austin. Many entities—government agencies, community-action groups and special-interest organizations—know how many people use what kinds of data they have available. "One of the things that should come out of this project would be to research these areas where people do accumulate data on what it is people need to know," he said.

Some computers already track system use and could help determine perceived citizen value of information by logging what material is used and what is not. Where computer data aren't readily available, researchers might analyze phone logs or survey people who use the services. Equally important is uncovering what kinds of information people seek but can't get. Defining demand for different kinds of information might be difficult and costly, however, as many organizations limit records to the requests they have filled.

Assuming that existing data might provide a basis for determining needs, Jorge Schement, a professor of communications at Rutgers University, urged evaluating information consumption patterns over time. "When we take slice-of-life studies that ask people what their information needs are, we tend to see them apart from time," he said. "While that can have some value, the time period is also important to include because of how patterns evolve over time."

To understand and provide for all segments of society, researchers will have to survey the needs of communities that are underserved by research-gathering agencies, such as poor rural areas. They must also consider the needs of individually disadvantaged users. Everette E. Dennis, executive director of The Freedom Forum Media Studies Center, pointed to the need for "a systematic look at all of those groups and individuals somewhere out of the loop, whether they are hearing impaired, visually impaired, illiterate or mentally retarded.

"There is a whole group of people, some of whom will have very great difficulty ever being part of this without some kind of technical assistance—the severely retarded, for example," Dennis said. "And that becomes a fairly substantial part of the population, once you begin to involve particularly those people who have marginal literacy of one kind or another."

Neglecting the needs of minorities and others who may be underserved would only exacerbate their disenfranchisement from the information marketplace, said Julius Barnathan, senior vice president for technology and strategic planning of Capital Cities/ABC, Inc.

"There's no concern for the minority, for the poor, for the people who are disenfranchised, for the people who live in rural communities," he said. "We find that education and illiteracy are getting worse, not better. So we need an information system to do one thing: educate. We've got to educate people so they can use these devices."

Deciding what kinds of information are needed for the public good also raises delicate market issues. For example, where should the line be drawn in providing information about energy efficiency? How-to information might be a public service, but the more specific the information, the more it might become advertising for products and services—which raises a related question: Where in an information service would advertising appear, considering that advertising does provide important information that people use?

Greater civic involvement might be another by-product of a citizens' information service, making it easier for more people to participate in debates and votes on pressing societal issues, including elections, said Alice Tait, associate professor of journalism at Butler University.

"The amount of time and energy that we consume simply working often precludes understanding and participation in the political process," Tait said. "Could we use such an information service to make available much more useful political information than what people get from current campaigns? Could people vote using this service?" Beyond elections, more people might take part through the interactive capabilities of the information system in public debates and decision making on a wide range of social issues, from education to health care to crime prevention.

LOCALIZING INFORMATION SERVICES

Most people are more concerned with local issues that affect them directly, instead of national or global matters, Kendall Guthrie pointed out. That being the case, she said, describing any new public information system as a "national information service" is not only misleading, since many of the available services would be local, but might put people off.

"I'm concerned a little bit, calling this a 'national information service,' which I think puts the perspective on national issues," Guthrie said. "I like to call them 'community information services.' People's communities are concentric circles—local and statewide and national. People want to know what their neighbors are doing in the PTA, what their friends think the best local restaurants or day-care centers are. In other words, we don't start out with the world; we start out with smaller areas and then gradually start connecting those up."

Susan Hadden, a professor of political science at the LBJ School of Public Affairs at the University of Texas at Austin, agreed, pointing out the practical side of local information. "If you're looking for the name of a plumber, you don't want one who lives 1,500 miles away, because presumably, you need him physically to arrive at your house and fix the thing."

The success of any such electronic information marketplace will depend on how useful people find the service, and whether they find there what they need. A related question is technological accessibility: "The services we're talking about ought to be just as accessible in Manhattan, Kansas, as they are on Manhattan Island," George Heilmeier said.

The question of local access and local services is also linked to reversing the trend in mass media toward metropolitan centralization. "Rural areas are increasingly underserved by mass media with a local interest," asserted Frederick Williams of the University of Texas at Austin.

"The local radio station maybe has a preacher on Sunday, but most of the time they roll these automated tapes of music and things like that," he continued. "The weekly paper is either not in existence anymore, or it's so narrow that it doesn't really cover anything of substance. People now get television, but it's brought in from another city. So there's no community information system." An answer might lie in the electronic information service.

Other panelists also objected to calling the proposed service a "national information service," saying that beyond failing to denote the local aspects of the system, it also implies information of a political nature. "Citizen," a suggested alternative, was also dismissed because of political ramifications—What about non-citizen usage?—and because it neglects the consumer element of such a service. "Information" was viewed skeptically because it implies one-way dissemination, shortchanging the service's interactive elements.

The fact that words often connote different meanings to different people, depending on their location, profession, and socioeconomic background, only serves to complicate the issue. Attempts to rename the information service yielded clumsy hybrids like "Nationwide Community Information and Communication Service," which, understandably, satisfied no one.

THE QUESTION OF ACCESS

But information, whatever its potential value, is worthless to those who cannot obtain it. If this comprehensive, new information medium becomes the means by which individuals and society communicate, then the question of information haves and have-nots becomes more acute, and the right of equal access becomes fundamental.

"I see us starting out by delivering services that existing institutions have shown that we already need, and doing it more effectively and cheaply," said Susan Hadden. "People have rights to get certain information, but often don't have the first idea how to get it."

Jannette Dates agreed, stressing society's interest in making information available to all and warning that without close attention to the question of access, existing information gaps could widen, translating into opportunity gaps. "In a nation built on the assumption of an informed citizenry, it's important for people to have access to the information that can assist in improving their circumstances," Dates said.

"It's been demonstrated, however, that increases in the flow of information usually widen the gap in knowledge and power between the people who have and the people who do not have," she continued. "Information systems, in my judgment, tend to distribute products in a form that is most familiar to users with more education—those who are already information-rich."

The paradox is that those who may most need information may at the same time be unaware that it exists, or of how they might benefit. "The information-poor often see little utility in much of the information that is available in society," Dates said. "Its very complexity is a barrier. Oftentimes they fail to see how use of information systems might be instrumental in improving their circumstances. I believe that one of the most important issues in the development of a national information service must be to help people perceive the service as important to their present circumstance."

A related concern is that some kinds of public information—for example, proposals to restructure utility rates—can become inaccessible or indecipherable because the structure of the information system permits interested parties to hide or bury it.

Kendall Guthrie cited an example: "I interviewed every single person in New Jersey who had ever made a right-to-know request between the passage of the New Jersey law and the passage of the federal law," she said. "And the problem that they all said was, 'We know we have this right, we try to exercise it. When we call, we get the run-around. They say they'll mail us the data. They don't mail it.' There was clear frustration among people."

If a citizens' information service is to be truly universally accessible, information must also reach the poor. In a country where many people cannot afford televisions, telephones or even radios, information-system terminals may well be out of reach of those disadvantaged and disenfranchised individuals who may well have the greatest need.

"To reach the poor, the service will need to make such a system available in publicly accessible places," Dates said. "Schools, for example, or libraries, shopping malls, the post office—places where poor Americans can be encouraged to use information systems to help them meet their needs for education, for medical care, jobs, skills training and the like."

Guthrie concurred. "Public libraries are increasingly trying to see themselves as public information utilities and not just book warehouses," she said. "A number of public libraries are now putting their card catalogues on-line and letting people access them from far away, or building databases of local information, everything from community services to all the kinds of things we're talking about and putting them in public libraries."

To minimize technological barriers to information, Guthrie suggested the mechanisms for accessing material be as simple as possible. "I would suggest the way to ensure universal access is to go for a very low-end technology," she said.

"Start out with something as simple as some audio text, with the menus printed in the front of the telephone books, so that 90 % of the American public could use this."

But Roger Fidler, a former Center fellow and director of new media development for Knight-Ridder Inc., disagreed. User-friendly graphics and audio software will be necessary both to attract people to the system and to teach them how to use it. "I don't think anybody is going to spend time accessing information by keying in information," he argued. "For any information medium to work, it's going to have to be predominantly visual and oral."

Evaluating the accessibility of technology that is still undeveloped involves much speculation. More predictable, however, is the direct relationship between sophisticated, user-friendly technology and cost. George Heilmeier, president of Bellcore, described the dilemma. "Given the diverse populations that will have to use this, the lower you go on the education scale, the more sophisticated the technology's going to have to be," he said. "So there's a flip here: When we talk about ease of access and simplicity, we have to understand that it's going to cost more to make information access easy."

Learning to use the emerging technology may be a generational issue, pointed out media policy analyst Benjamin Compaine. "Children are in greater synch with technological developments than their parents," he said. "There's a generation of kids who have been playing video games for a number of years, who have had access to Apple computers, for whom doing the VCR is a piece of cake," he said. "They don't have a lot of the same problems we have, about access to anything. There is a generation of kids coming on, including kids from the poorest families, who are much more glib with a lot of this technology than some of us older fogies."

But Anne W. Branscomb, research associate in the Program on Information Resources Policy at Harvard University, suggested that this generation may not really be much more skilled at using sophisticated new technology. "If you look at the cash register that's used in McDonald's, for instance, they clearly don't think their employees can use anything very sophisticated. It's all little icons," she said.

"The technology does provide instrumentation for people with a fairly low level of literacy," Branscomb said. "But I think you're talking at cross-purposes if you say on the one hand, 'The problem's going to solve itself,' and say on the other, 'We need this very expensive, voice-activated, visual system that doesn't use keyboarding or print literacy at all.'"

New York Newsday's Fred Tuccillo agreed, pointing out that education is at the root of the dual question of access and ability to use the technology. "It seems to me that educators, particularly those on the primary and secondary level, would have a lot to contribute here," Tuccillo said. "We're talking about the eventual audience for these services, and we're also talking about an institution in society that can play a great role in helping to shape the services and prepare citizens to use them."

Others agreed: In the same way that teachers take students to libraries to learn how to navigate card catalogues, they must teach students to use a national information service. This means teachers and others on whom citizens rely for guidance must understand the system first.

"We're talking about moving information from large providers to individual citizens and not spending enough time talking about the middle people—the social workers, police officers, school counselors and all of the other information brokers," Everette Dennis said. "I think we must make it a very high priority to see that those kinds of individuals can be better informed. We must have information literacy, and it has to be taught somewhere in the basic grades."

Reliance on information brokers may subside somewhat as individuals become more accustomed to the system's technology, uses, and organization. On the other hand, given the world's vast and constantly growing amount of information, brokers may become even more valued as guides to lead users through their informational expeditions.

THE THORNY QUESTION OF POLICY FORMATION

Fulfilling needs, ensuring access and promoting education will depend on policy. Historically, U.S. information policy has been vague at best.

Stuart Brotman, a senior fellow in the Annenberg Washington Program, said defining information policy was a perennial problem during his tenure in the National Telecommunications and Information Administration. "I think the puzzling question for all of us was, What is information policy?" he said. "In the 12 or 13 years that the agency has been in existence, it's been very difficult for people in government to understand what information policy is and how government can somehow organize it." As a result, policy has tended to be reactive rather than proactive, Brotman said, emerging once a medium was in place and responding to circumstances after the fact.

Anticipating a citizens' information system based on possible future technologies presents the opportunity to develop policy that addresses the needs of all users, as well as those who have a financial or political stake in shaping the new medium.

Donna Lampert, an attorney with the policy division of the Federal Communications Commission's Common Carrier Bureau, suggested that the fundamental goal of a national information service policy should remain consistent with the goals of existing media policies. "That has been the First Amendment goal, to foster a diversity of access of information sources, not a diversity of programming or information but of different sources," she said. "That's the premise under which the First Amendment operates: The more information we get, the more likely we're able to reach right conclusions and fulfill the goals we have for a democratic society."

Many policy questions affecting telecommunications and media industries are already under scrutiny. Should a public information service be regulated as a public utility? Can a carrier be a provider? How to protect against monopoly?

Is redundancy a necessary safeguard or a waste of resource? Suggested answers to these and other questions can be heard in courtrooms, Congress and public forums across the country.

Much of what becomes policy will be shaped by how citizens and public servants ultimately regard information. Viewed in the context of marketplace-of-ideas theory, which helped form the First Amendment, the policy issue becomes one of ensuring the greatest possible and most wide-open and robust debate through any information service.

"The growing discrepancy between those with access to information-age service and those without can well be interpreted as a threat to First Amendment guarantees of our Constitution," Frederick Williams asserted. "Those who cannot communicate via the new networks or who cannot have easy access to its information services are being denied participation in our nation's newest medium."

This raises a question of definition: The First Amendment was intended both to guarantee diverse voices the right to speak and to provide society with a wide variety of views on which to base societal decisions. Providing for an equitable distribution of limited channel space in the electronic information marketplace will inevitably become a political issue, the panelists said.

"If what we're striving for is some information source or a series of sources that are going to be regarded as authoritative, then you begin to get into the First Amendment question," said John Pavlik, the Center's associate director for technology studies, "because it suggests that you will design a system that will make it easier for these sources to disseminate their information. They will have some kind of competitive advantage, perhaps, over everyone else in the marketplace of information and ideas. That's where I think there's a First Amendment consideration."

Policy must also be fashioned to determine what information people are entitled to receive. For example, should information about rape and incest counseling include a list of abortion services? Or can the list be included with pregnancy and birth-control information? Can abortion services be listed separately and advertised or should they be excluded completely? Ongoing debate over abortion counseling at federally funded agencies indicates this question could well continue in the electronic information context, depending on system funding sources.

"I know, as a journalist, that even a grocery list is not going to be deemed by everybody to be politically neutral," Fred Tuccillo said. "Any information, in whatever form, will somehow offend the sensibilities of someone."

COMPETITIVE FAIRNESS CONCERNS

From the providers' point of view, limited space in a citizens' information system portends issues of competitive fairness. Existing media will have some legitimate concerns about the impact of a new information system on information sources already in place, pointed out James Hoge, a senior fellow at the Center and former publisher of the New York *Daily News*. "I think a key public issue concerns how we preserve the existence of news-based products that are commercially viable as part of the information system of the country," he said. "That is a different issue from preservation of a particular kind of medium—say, news presented as ink on paper."

The cost of producing newspapers is becoming prohibitively expensive, Hoge said. As a result, newspapers face difficulties both in maintaining a commercial base and in surviving as a mass communication medium. The medium must evolve to exploit and expand its strengths, he said, and an electronic medium for disseminating information offers new opportunities for newspapers.

"The newspaper business, on the product side, is of two different parts: one gathering news and the other printing," Hoge said. "Too much emphasis, it seems to me, is placed on the printing side rather than on maximizing the historic talents and resources that have been brought together in gathering information. I think that any low-cost, easy-distribution, high-access information system would be the best place to assure news-based products in the future."

Tuccillo agreed. "Newspapers are well aware of their own tremendous information resources and of the fact that, in the future, they're going to have to derive more and more of their revenue from those sources," he said. "But the basic issue that the newspaper industry has not confronted is the question of the time horizon on which they base their investment and policy planning. They are so preoccupied with the immediate threat posed by the telephone companies that they have dedicated themselves to meeting that on the political front and disregarding the need for a long-term financial investment, which at the moment they're too strapped to make anyway."

EXPRESSION, PRIVACY AND RESPONSIBILITY

Another key policy concern, panelists said, will involve both free-expression issues and privacy protection of system users. If the service is an interactive public system to which access is a fundamental right, all citizens must be assured equal opportunity to express opinions and ideas; the debate over protection of ideas that some find to be hateful or obscene will continue in the electronic information marketplace.

Universal access to an interactive system also raises questions about privacy, anonymity, responsibility and security. The panelists grappled with a number of these issues: Should users be required to identify themselves to gain access to the system? If so, what happens to the computerized lists of users of certain information? Will users be able to protect themselves, for example, from telemarketers and mass mailers, who could hawk services and products based on the cross-indexed call logs of the computerized information service?

Of more serious consequence is the question of what happens to the list of people who access potentially compromising information about, for example, AIDS or mental health treatment. The possibility that those who most need help available through a national information system might not use it for fear of identifying themselves would exacerbate the gap between the information haves and have-nots. Even an anonymous identifier such as a social security number or driver's license number can be traced to the owner.

Anonymous access also poses dangers. "There are certain reasons for anonymity," acknowledged Lawrence Grossman, former president of NBC News and PBS, "but anonymity also breeds irresponsibility."

Under existing systems, publishers, broadcasters, and other information providers are held responsible for their message content. But the question of responsibility becomes confused when the distinction between provider and user is blurred or indistinguishable on an interactive public system in which users may remain anonymous.

Even with identifiable providers, the reliability of information is always questionable. Harvard's Anne Branscomb suggested that policy will be needed to address the issue of authenticity in the collection of information. "Even in the census, we get much less data than we need to do reliable demographic analysis because people are apprehensive about disclosing information about themselves," she said. "If you were a migrant Mexican, for instance, here illegally, you wouldn't want to disclose that. And yet, we need the statistics to know how many children are appearing in public schools or using other social services."

Gatekeepers in an electronic information system will operate under the same constraints and influences that confront gatekeepers in existing media. When the gatekeepers are known entities—the *New York Times*, for instance, ABC News or the *National Enquirer*—the public may evaluate reliability based on the source's track record. Gauging the reliability of information on an interactive information system in which users could be anonymous would be much more difficult.

Many people already rely on human intermediaries, trustworthy friends or experts who guide them to reliable sources of information and help them form opinions. The blurring of the line between source and user in an interactive electronic information system might provide loopholes for advertising to masquerade as objective information. Such loopholes already exist, some panelists said.

Susan Hadden cited a Texas example. "In the Austin phone book, in addition to the *Yellow Pages* ™, we also have some shiny pages that offer single telephone numbers and then a lot of four-digit codes that access health information, environmental information and so on," she said. "In the health information, for example, if you dial a four-digit code for breast cancer, you get a recorded message that tells you the possible treatments. But if you dial the four-digit code for plastic surgery, you actually get an ad that was paid for by a particular plastic surgeon. It's the same format, not distinguishing between advertising and public service. "I find this very troubling," Hadden said, "particularly for those people who are least able to distinguish that something is an ad from something that isn't."

Security is yet another concern. Transmitting information electronically presents all kinds of potential security problems. If access is anonymous, information might be in danger of being altered or stolen. The question remains how to protect proprietary information that is deemed to be useful to the public.

This raises issues pertaining to intellectual property law. If information is disseminated in a public forum, it might well be considered in the public domain, opening the possibility that it could be repackaged and marketed elsewhere. Without the prospect of compensation for the original "owner" of the information, there is little incentive to share it in the first place, thus limiting the goods in the electronic marketplace of ideas. Information that is protected, ownable and marketable for profit would effectively be inaccessible to the poor.

Further, if information empowers, it also can be used to damage. "When we're talking about legal responsibility," said Annenberg's Stuart Brotman, "we're talking ultimately of people seeking redress for all sorts of problems that may arise in utilizing this network, whether it's incorrect information or emotional distress caused by receiving certain information or transmitting it. I think we need to look at our tort system and whether this would somehow create major cracks in that system, or whether that system could accommodate it, or whether we are essentially going to create a litigation battle."

However thoroughly scholars and policymakers manage to anticipate the legal ramifications of a national information service, any new policy will likely serve as the basis for years of judicial decisions.

EXISTING POLICY MODELS

As Donna Lampert of the FCC had suggested, existing policy provides guidelines for new electronic information policy Bellcore's George Heilmeier said.

"I think a lot of the needed building blocks for policy are already out there," Heilmeier said. "For example, pieces of legislation like the 1991 High Performance Computing Act, the Copyright Act, the Freedom of Information Act, and Title IV of the Americans with Disabilities Act—which expands the notion of universal service—can at least give us a better understanding of where legislatures may move in this direction. The Television Decoder Circuitry Act talks about essentially installing technology and they having the market follow it. We could probably list a half dozen to a dozen federal acts on which new policy could be based."

At the same time, consideration of previous policy must anticipate the unique features of a new medium, Lampert pointed out. "I think one of the things we need to do when we develop policy is to recognize that, as technologies converge, we can't draw the same distinctions we have in the past, based upon particular kinds of service," she said. Existing video, audio and print media are segmented, she said, but a medium that combines all three will blur many of the distinctions that have served as boundaries for previous policy.

"I think it's a slippery slope to focus too much on a model act," Stuart Brotman said. "I don't think Congress is much interested in hearing a notion of an act from external people who come in and do work policymakers feel that they are more charged to do. To the extent the vision is couched in some other way, as opposed to one of these Ten Commandments-type documents, I think it would probably be better perceived."

But George Heilmeier disagreed, pointing out the intent and benefits of creating a model bill. "I think what you do, in drafting a bill, is not with the expectation that somebody's going to introduce it as it is," he said. "But you're advancing the discussion to a different level, to a nuts-and-bolts consideration of specific subjects, specific clauses, specific aspects that are questionable or debatable, or for which people want a substitute. You're forcing a confrontation with specifics, rather than with the generalities of public policy objectives."

Although offering policymakers specifics might be helpful, Susan Hadden said, the overall policy itself should take a broader perspective. "It seems to me that one of the nice things about the Communications Act of 1934 was that it didn't lay out in a lot of detail what was supposed to happen," she said. "Practically all of the policy implications that we're talking about now derive from the tiny little preamble. What we really need from Congress isn't a law exactly, so much as an agreement on some kind of vision that can help the FCC and state regulatory bodies and all these diverse groups have a sense of where we're going."

Lawrence Grossman agreed: Beyond the mechanics, it is essential that any policy governing a new information system enunciate basic philosophical principles, he said. "We need to affirm a basic principle of fostering the things that we value in our society—education, culture, major information," he said.

"There has to be a public service component to this. It will never be the driving force of all of it, but I would establish a fundamental principle that, in addition to whatever commercial uses will drive this policy, there must be a public service component preserved."

COSTS, FUNDING, AND PROFIT QUESTIONS

No practical guidelines for an information service policy can ignore the issue of cost, at least in general terms. Indeed, any theoretical discussion of technical feasibility, information needs or user access and education is ultimately grounded by the question: Who's going to pay for all this?

One might begin with the common-sense response: Those who profit from it. But complications arise immediately. If a citizens' information service benefits society, then society must bear at least part of the cost. But questions of taxation on local, state, or national levels, coupled with issues of local control, quickly become complex and political. Another possibility would be for commercial players in the information marketplace to shoulder the cost burden. Presumably media, telecommunications, and information services industries, profiting from access to contact with millions of consumers, would pay for such access.

But since the system would consist of a mix of public and commercial services, funding will probably come from various sources. Determining how much from whom and by which method may involve forecasting a profit ratio that includes benefits less easily appraised than straight dollars.

Assuming that a unified information service consists of a single carrier bearing many providers, ownership of that carrier system promises to be profitable. Already telephone, broadcast, and cable companies are battling for control; companies developing alternative transmission media—such as Direct Broadcast Satellite and Multipoint Distribution Service—may soon join the fray. Wherever ownership finally falls, it is arguable that the controlling entity should finance the carrier system from which it will profit.

The issue then becomes how to ensure low-cost or free public access in a privately owned system. Historically, this involves a policy of government regulation that sets how much a company pays for the right to run a regulated monopoly. The goal is to secure equitable public access without destroying the owner's profit incentive. In the cable industry, for example, local franchises are auctioned off, based in part on the provider's agreement to reserve a percentage of channels for public access. The success of past regulatory policies in achieving this goal remains arguable, but they may nonetheless serve as the basis for policy on the funding structure of a new electronic information service.

FINANCING THE SYSTEM

Once the carrier system is in place, support for the information service may come from vendors. By the same reasoning that trucking companies, for example, pay usage taxes for conducting business on public highways, commercial information providers could pay similarly constructed fees. This money, in turn, could be redirected to fund the public-access component of the system.

Frederick Williams pointed to the Dade County Public Information System, which offers access to real estate records, tax information and other county information on-line. Commercial user fees provide enough income to support the system's hardware and software needs as well as supporting services such as community calendars, Williams said.

"Dade County is actually just the first to do this," New York City policy analyst Kendall Guthrie said, pointing out that other municipalities nationwide are moving in a similar direction. Automated real estate and other records for which there is a demand from commercial users generate fees to pay for services such as health information.

"There has proved to be a real market of convenience for lawyers and real estate and title companies not to have to send somebody down to City Hall to get that information," Guthrie said. "And they are more than willing to pay for the convenience of being able to get it in their home."

Another funding method might be to treat channels as franchises that can be auctioned off, Lawrence Grossman said, with franchise fees supporting various public service uses, "just the way Weyerhauser is charged for logging in the public forests, and that money goes to replenish the public lands."

But these kinds of funding mechanisms involve the kinds of cross-subsidies that "we spent the last 10 years getting rid of in the telecommunications system," Anne Branscomb objected. "I mean, we broke up AT&T because of the cost of it. The economists said, 'Oh, this is the wrong way to go. Everything has to be paid according to its cost and the right way to do it is, if you have a service for which people cannot pay, it's got to come from the public coffers as a direct tax.'"

But Susan Hadden suggested reconsideration of cross-subsidies, which exist in other systems, including the postal service, broadcast media and corporately owned public works. "We have a long tradition in the United States of cross-subsidizing lots of things," she said. "Research into the changes in ideological and legal positions that led to our present belief regarding cross-subsidizing would be very useful.

"I also think we have degraded the notion of public in present political rhetoric and that we need to reinstate the notion of public and public good and public interest and public utilities and common carriage," Hadden said. "The fact that we have things in common is what makes us a country."

COOPERATIVE EFFORTS

Although it may seem farfetched in the current combative climate surrounding media industry struggles for market control, the panelists said, it may profit everyone to work these details out together.

"It strikes me that instead of the major business players dividing up into camps, there's a lot to be gained by newspapers and telcos looking at how they can cooperate, as opposed to how they're going to outgun each other," George Heilmeier said. "And there are plenty of opportunities for cable companies and telcos to behave in a cooperative manner rather than as sworn enemies."

Fred Tuccillo acknowledged the potential benefits of cross-industry cooperation based on what he sees happening already within newspapers. "Having sat people down at *Newsday* who don't normally sit down with each other—people from editorial, people from the business side and people from the technological side, such as it exists at a newspaper," he said, "one of the things we're beginning to see is that there are bridges across these questions that aren't going to get crossed unless the people who put out the newspaper can articulate things they might like to do, services they might like to provide, in the presence of people who can immediately say, 'Well, wait a minute, this technology can help you do that,' or, 'There's somebody developing a new transmission method that may enable you to do even more than what you've described.' That kind of chemistry is just now beginning to develop at newspapers."

But Rutgers' Jorge Schement suggested that media industries alone will not be able to develop and implement this new system. "If one looks at the history of infrastructures in the U.S., no dialogue has ever led to a new infrastructure unless business was talking to government," he said. "And since World War II, no major infrastructure changes have ever occurred without the military. In this case, we're talking about a number of different players of varying size and importance.

"It's important for them to be represented in some way so they can at least confront and listen to each other, but also confront and listen to government," Schement said. "And the military presenting their interest in this, or being present in order to hear what's happening on the other side, might be valuable in terms of affecting policy."

Presumably such a discussion would also include consumers, now represented in the debate by various groups voicing diverse goals. Among them are those who deny the need for a comprehensive information service on a national or global scale in the first place. "We really have a big tension here," said Donna Lampert, "because on the one hand, we're making the argument that these are the people that are going to benefit. But there are people who purport to represent them, who say, 'Hey, we don't need it. Everything is fine, back off, we don't want to pay for this.'"

Dissent also has spread among groups who claim to support a national information service, Susan Hadden said. "There are four or five groups now that are talking about the utility of the broadband network," she said. "But one of them is focused on public interest television kinds of concerns; one of them is concerned about technologies; some of them are known to be closely associated with specific industrial stakeholders. And so on."

To influence policy formation, consumers will have to reconcile disparate goals and promote agreed-upon priorities in a unified voice, the panelists agreed. But progressive discussion among any of the concerned parties is tempered by mutual suspicion, John Carey of Greystone Communications said. "It strikes me that one of the things that will have to happen for this network to form will be trust among groups that don't trust each other," he said.

"Many people don't trust the phone company. Many people don't trust cable companies. No one seems to trust the government. That's clearly a problem," Carey said. "What we need to do, if this is going to form, is to build trust and cooperation among groups who essentially have not had a lot of trust for each other."

Considering that many of the interested parties make their livings by supporting a particular point of view, that trust will be hard won, the panelists concluded. Rather than letting isolated skirmishes continue to play themselves out in scattered locations and forums, perhaps some accord could be achieved by bringing the diverse would-be players together with consumers to define areas of unity and dissent. The Freedom Forum Media Studies Center's national conference may present the opportunity for just such a meeting of minds, and an important step forward for a broad-based citizens' information service to serve a diverse population of diverse informational needs.

ADDENDUM: THE ADVANCED INTELLIGENT NETWORK

When the roundtable discussion addressed technology, Benjamin Compaine suggested, "Any technology that's going to be commercially available 5 or 8 years from now already exists. What you want to do is say, 'OK, what is some of the state-of-the-art stuff that's in the labs today, that people are playing with up at MIT or at GE or at Bellcore? What's there that might fit into the things that will make information access easier?'" This question was responded to by Bellcore President George Heilmeier, who introduced the roundtable participants to the advanced intelligent network (AIN), a network architecture currently in development at Bellcore laboratories. Generating interest and questions, Heilmeier responded to various issues raised by the panelists regarding AIN. What follows is an edited transcript of Heilmeier's presentation.

The telcos are involved in a very big change in philosophy now, called AIN—the advanced intelligent network. AIN is rolling out now and will begin having a major impact in the next 2 to 3 years. It will be built for the rest of this decade.

There's a fundamental paradigm shift in AIN. It used to be that if you wanted to create a new service, the only people that could do it were the telephone companies. And the telephone companies would go to the switch vendors—like AT&T—and they'd say, "We'd like you to change this switch to support this service."

Well, the switch vendor immediately says, "I've got to talk to all the other companies to see if there's a real market for this." And so, your idea is socialized immediately.

Then, if the switch vendor decides that it's a good thing to do, it probably takes another 2 1/2 years, and lots of money, to build the switch software. That's the way the system works today.

The advanced intelligent network enables service creation and control to be programmable by the user. For the first time, the telco can create new services itself without having to seek major switch software changes. And that has certain implications.

Number one, it means that third parties and customers are going to be able to customize their own services. You could never do that before—it was too expensive, too time-consuming and too sophisticated.

When you make service creation and control programmable by the user, you get yourself out of essentially having to revise a switch software system that may have 10 to 15 million lines of code in it. This is done by developing service-independent building blocks that are linked together by service logic that exchanges information with the switch.

When the concept of AIN was first formulated, some people thought, "Well, we have this great service-creation environment, but only the telcos should create new services."

On further reflection, they said, "Look, that simply isn't going to work. We don't understand all the application domains. The smartest thing we can do is to make this environment available to customers and third parties."

And that's exactly what's going to happen. AIN will provide the environment that enables third parties and customers, as well as the telcos, to define and customize services. So new services will be easier to try. If they work, you can grow them throughout the environment.

So this is a major shift, testing whether there are enough people out there who are interested in customizing their own services, and then letting them do it themselves. Certainly there are lots of technology and policy issues. But the telcos have crossed the bridge, saying in effect, "We alone cannot invent all the interesting services out there."

AIN is different from, say, Integrated Services Digital Network (ISDN) which is a transport architecture that gives you a "dial-tone for data" capability. ISDN essentially puts in the hands of people who are interested in transporting data the ability to do more efficiently that over the existing phone system.

That's a lot different from giving people the ability to

actually invoke switch features and customize services. AIN permits you to define new services that use the network's smarts.

Let's say I want to have a national number, but I want it to be local. For example, a person who calls for a pizza hopes to get it in 30 minutes. Well, it's not going to come from someplace 100 miles away. AIN could route the call to the closest site. Defining those kinds of services would very much be in the spirit of AIN.

Likewise, the platform which the advanced intelligent network provides could be the beginning of localized community information services, which could grow into the national information service we've been discussing.

Also, AIN has the flexibility to accommodate new technology, a generation or two hence, while at the same time not confronting us with the policy question of eliminating everything that's already in place. The flexibility is there because the principles of AIN are principles that are implemented in software.

You can argue about the size of software releases and the complexity of generating large software systems. But we're not talking about essentially digging up everything that's already in the ground and starting over again in the future.

All of the Regional Bell Operating Companies, in one form or another, are involved in AIN. It will roll out faster in some places than others, but initial advanced intelligent network capabilities will be widely available within the next 2 to 3 years.

ACKNOWLEDGMENT: An earlier version of this chapter was published in 1992 as a a Freedom Forum Media Studies Center briefing paper, edited by Edward C. Pease, the Center's associate director for Publications, and Leo Fitzpatrick.

APPENDIX
Roundtable participants and their affiliations:

JULIUS BARNATHAN, senior vice president for technology and strategic planning, Capital Cities/ABC, Inc.
LEO BOGART, former executive director, Newspaper Advertising Bureau.
ANNE W. BRANSCOMB, research associate, Program on Information Resources Policy, Harvard University.
STUART BROTMAN, senior fellow, Annenberg Washington Program, Northwestern University.
JOHN CAREY, director, Greystone Communications.
BENJAMIN COMPAINE, president, Samara Associates.
JANNETTE DATES, associate dean, School of Communications, Howard University.
EVERETTE E. DENNIS, executive director, The Freedom Forum Media Studies Center.
ROGER FIDLER, director, New Media Development, Knight-Ridder, Inc.
PATRICK GARRY, research fellow, The Freedom Forum Media Studies Center.
LAWRENCE GROSSMAN, former president, NBC News, PBS.
K. KENDALL GUTHRIE, policy analyst, Department of Telecommunications and Energy, City of New York.
SUSAN HADDEN, professor of political science, University of Texas at Austin.
GEORGE HEILMEIER, president, Bellcore.
JAMES HOGE, former publisher, New York *Daily News.*
DONNA LAMPERT, attorney, FCC Common Carrier Bureau—Policy Division.
JOHN V. PAVLIK, associate director for research and technology studies, The Freedom Forum Media Studies Center.
FRED TUCCILLO, assistant to the editor, *New York Newsday.*
JORGE SCHEMENT, professor of communications, Rutgers University.
ALICE TAIT, associate professor of journalism, Butler University.
MARK A. THALHIMER, technology manager, The Freedom Forum Media Studies Center.
FREDERICK WILLIAMS, professor of communications, University of Texas at Austin.

Part II

Citizen Information Services and the Public Interest

A focus on the citizen loomed important for two reasons. First, some of the most recent innovations in electronic information services are those locally developed to serve city or country needs (i.e., specific citizen populations). Research such as that reviewed in chapter 5 reveals that in addition to information, users also seek transactional and messaging services in their communities. Second, if we make good on our emphasis on the citizen-user of services, there are important policy issues such as that addressed in chapter 6 on access, involvement, and First Amendment.

Chapter 5

LESSONS FROM PUBLIC AND NONPROFIT SERVICES

BY WILLIAM H. DUTTON

Innovations in electronic service delivery in the public and not-for-profit sector provide a perspective on the public's use of electronic information services. This chapter surveys innovations in electronic service delivery, focusing on lessons that can be learned from these experiments. If emerging trends persist, strategies for supporting the public's information and communication needs should focus on facilitating the migration of information and services to electronic media of all types rather than promoting any single medium for reaching the public with information perceived to be in the public interest. At the same time, the providers of public services must insure access to public facilities and support for groups that will be electronically disadvantaged, if not disenfranchised, without vehicles and on-ramps to the many "information highways" criss-crossing the United States. *William H. Dutton is national director of the U.K. Programme on Information and Communication Technologies (PICT) at Brunel, The University of West London. He is on leave as a professor of the Annenberg School for Communication, at the University of Southern California.*

ISSUES OF CITIZEN WANTS AND NEEDS

Visions of a national electronic network often lack compelling evidence about the public's need or interest in this new age information highway. What services would be delivered over this superhighway of the information age? Will the public interest truly be served by its development?

Answers to these questions from the proponents of new telecommunication infrastructures are often quite vague. Frequently, the answers are reminiscent of overly optimistic forecasts of earlier decades in which many politicians, scholars, and journalists also foresaw a wired nation. Forecasters have had a poor track record in anticipating the public's response to such communication and information technologies as interactive cable television, the videophone, and videotex services. Few decision makers are any longer willing to invest on the basis of optimistic scenarios about the public's enthusiasm for new electronic services. An increasing number of telecommunication experts simply appeal to faith in human ingenuity (Lucky, 1989). They argue that the future of information services is unknown, but that once new technologies are in place, the services will soon be invented—just as in *A Field of Dreams*—"They will come."

It seems, however, that a case for electronic services might be emerging from a variety of initiatives being undertaken by American state and local governments, which have long been viewed as natural laboratories for experimenting with innovations in public services. Since the late 1980s, several dozen federal, state, and local government agencies as well as a number of quasi-governmental, not-for-profit agencies have been experimenting with the provision of electronic information services to the public. Their experiences provide one of the only demonstrated perspectives available on the kinds of public needs and interests that could be served by electronic information services.

These projects underscore the diverse range of applications, technologies, and tasks that will define the future of electronic service delivery. They also provide an empirical basis for challenging commonly held assumptions about the future of public information networks and services. Taken together, these innovations in electronic services might indeed foreshadow a long-term trend toward the more direct provision of information services to citizens via broadcast, telecommunication, and other electronic media that are anchored in identifiable needs of the public. At the same time, these projects suggest that some visions of the future of electronic services are based on an overly simplistic view of the public's interest in information services.

RECENT EXAMPLES OF PUBLIC AND NONPROFIT APPLICATIONS

A small but increasingly visible number of state and local government agencies as well as not-for-profit organizations are applying communication and information technology in ways designed to facilitate public access to public information and services. These innovations find expression in a variety of different types of applications across a range of functional areas. Although there is no master plan shaping these individual initiatives, there are several threads that tie these efforts together.

One thread is an effort to emulate and catch up with the private sector on the assumption that the modernization of communication and information systems will bring major technical benefits in the speed, accuracy, and efficiency of public services. Another is a more or less explicitly articulated vision of using electronic media to bring government closer to the public. But the long-term, indirect consequences of these developments might have a relevance to politics and governance, which goes beyond improving the efficiency of existing practices, to altering the way we do things—such as voting, polling, lobbying, communicating, and obtaining services.

Generally, innovative applications of communication and information technology fall within four general types of applications, described in Table 5.1; namely, broadcasting, transactions, access to public records, and interpersonal communications.

TABLE 5.1
Examples of Services Supporting Citizen Access

Communication Task:	Systems Employed:	Early Applications:	Selected Examples:
Broadcasting	Satellite-linked and local cable TV systems	Cablecasting of public meetings, hearings; neighborhood TV	C-SPAN, California Channel (Cal-SPAN); West Hartford, Connecticut
	Electronic bulletin board	Continually up-dated information on events, agendas, services	Santa Monica's PEN, Pasadena's PARIS/PALS
	Touch Screen, Multi-Media PC	Mulit-lingual kiosks	24-Hour City Hall, Hawaii Access, LA Project, LA County Library Pilot
Transactions	Touch Screen, Multi-Media PCs or kiosks	Apply for social services	Tulare Touch
		Renew/up-date driver's license	California DMV's Info/California
	Magnetic strips and smart cards	Electronic benefit transfers	New York City Food Stamp benefits
	Voice processing	Schedule inspections	Arlington County
	Electronic mail	Requests for services, completion of applications, licenses	Santa Monica's PEN
	Automated Teller Machines (Kiosks)	Welfare and medicaid transactions	Proposals at the state and federal levels
	Automatic vehicle identification systems	Automated assignment of tolls for use of roads	Under discussion in California

TABLE 5.1 continued

Access to Public Records	Audiotex, recorded messages	Answers to routine public inquiries	Hillsborough County's Fact Finder; Phone Phoenix
	Dial-up electronic bulletin boards	Public, government, community information	PARIS/PALS Pasadena; Santa Monica's PEN; New York City Board of Education's NYCENET
	Dial-up access to public databases	Access to land records by title companies	Proposed in Arlington County, Virginia
Interpersonal Communications	Expert systems to assist information providers	Assist information & referral providers in offering advice & referrals	Under development by INFO LINE in Los Angeles
	Voice mail, asyncronous voice communication	Parent-teacher exchanges	City of New York
		Inquires about teacher certification	State of Connecticut
	Facsimile	Application for building permits	City of Spokane
	Electronic mail	Citizen complaints, inquiries, requests	Santa Monica's PEN
	Computer conferencing & bullentin boards	Electronic forums on public issues	Santa Monica's PEN; NYCENET, FreeNet
		Electronic networking to support groups & voluntary organizations	Santa Monica's PEN, HandsNet, Community Link, Reference Point
	Audio and video conferencing	Arraignment; bond reviews; and other pretrial meetings	San Bernadino, CA; Dade County, FL, ...

Note: This table is adapted from Dutton, Guthrie, O'Connell and Wyer, (1991); and Dutton, (1992c).

First, the most prominent uses are aimed at broadcasting information for the general public. As cable systems have diffused, public agencies have more frequently used these channels for distributing government and other public information, such as by scrolling textual information or cablecasting live events, such as council meetings. Cablecasting of governmental affairs is becoming a common practice, but localities like West Hartford, Connecticut, have advanced the state of the art in this area by producing more sophisticated local news and public affairs programming and committing themselves to the concept of "neighborhood TV" (Dutton, Guthrie, O'Connell, & Wyer, 1991). Rather than simply pointing a camera at live events, these producers and growing ranks of volunteers seek to translate lengthy meetings into bite-size programs that capture the interest of the local audience. Old visions of community programming over cable are beginning to be more closely approximated in a few localities.

Beyond the local level, the success of the Congressional-Satellite Public Affairs Network (C-SPAN) has prompted state officials to emulate their programming (Westen & Givens, 1989). California has launched a system, which it calls the California-Satellite Public Affairs Network: The California Channel (CAL-SPAN).

State and local governments are also experimenting with using information kiosks, multi-media personal computer platforms, for distributing information to the public. These kiosks generally combine laser disc storage devices and micro-computers with a touch screen to allow people to access government and other public service information. They are being placed in libraries, shopping malls, recreation centers, and grocery stores to provide public information in the form of text, video, graphics, or sound, all in multiple languages to reach a more multicultural society. They are promoted as a means for bringing information closer to communities, not expecting citizens to travel to city hall or speak English.

In this area, Public Technology, Incorporated, in partnership with IBM has supported a "24-hour City Hall" project, which placed kiosks in over a dozen local jurisdictions. In 1992, immediately after the Los Angeles protests and riots, North Communications, a Los Angeles firm that has worked with IBM in developing kiosk systems, launched what it called the LA Project. North distributed a small number of kiosks throughout Los Angeles, designed to support South Central and neighboring Los Angeles residents in seeking phone numbers, public information, business loans, employment, and other assistance of possible value to recovering from the urban violence that followed the Rodney King verdicts.

A second category of examples are transactions services. The Department of Social Services in Tulare County, California, has developed a kiosk system called "Tulare Touch." Tulare Touch uses a multilingual, video and audio touch screen connected to an expert system. The department has 35 kiosks with a touch screen at each district office. Welfare clients step up to a kiosk, choose which language they wish to use, and are walked through a welfare application. Fraud screens provide checks of income or other information by linking with other files to verify the client's eligibility. The system approves or rejects the applicant at the time of application or gives the applicant a list of information needed for any future application. The project was aimed at reducing costs by using the capabilities of an expert system to minimize errors in determining eligibility and calculating payments.

In late October 1991, California's Department of Motor Vehicles (DMV) launched a 9-month pilot project called Info-California (Dutton et al., 1991). Fifteen information kiosks were initially placed in Sacramento and San Diego to pilot this experiment, permitting citizens to conduct a variety of transactions, such as registering an out-of-state vehicle. The system was developed by IBM and North Communications. This system was modeled after an automated teller machine.

One major thrust of the Info California project is to overcome the jurisdictional fragmentation of services and permit the public to access information at any level of government through a single facility. The director of California's Health and Welfare Data Center wishes to "provide a single face to government for the citizens of California" (Hanson, 1992). Various agencies in California, such as the Los Angeles County Library, which have had some success with stand-alone projects, are actively considering the use of the Info California system as a platform for the wider offering of their services.

Other state and local governments are using similar electronic systems for distributing benefits, such as food stamps, hoping to provide recipients with benefit cards that could be used much like a credit card in local groceries and supermarkets. Other government agencies are experimenting with smart cards as another approach to automated payment and billing.

Arlington County, Virginia, has implemented what is called the "parkulater," a device about the size of a pocket calculator. It uses a small computer chip and acts as electronic dollars for commuters parked at meters. The user types in the type of meter, places it in the car's front windshield, and the time runs until the driver leaves and turns it off. Drivers are billed for time they are actually parked and they do not need to have the correct change. By 1992, as many as a half dozen other jurisdictions were using this technology and already beginning to face problems with a lack of standards for parking fees across neighboring jurisdictions.

A third category of services provides access to public information and records. New technologies are being employed to provide an additional means for providing public access to information and records. The technologies employed range from audiotext, menu-driven telephone systems, facsimile, computer bulletin boards, and videotex, to multimedia personal computer kiosks. All are designed to supplement rather than replace more conventional systems for accessing public records over the phone or over the counter.

Voice mail systems are being used to answer routine

citizen inquiries. "Phone Phoenix" is one system offering prerecorded messages over the telephone as a means for answering many common, repeated questions and disseminating information, such as a schedule of current events. The system records menu selections made by the public, which also enables the city to determine what topics are of most interest to citizens. Some cities are using these systems to handle more specialized information needs. Dallas, Texas, uses audiotext to communicate with people requesting information about the status of building permits. Arlington County has implemented a form of audiotext that provides building contractors dial-in access for scheduling a building inspection.

Like many other telephone-based information and referral agencies, INFO LINE, a nonprofit California corporation, uses a menu system to route calls to the appropriate information specialists, such as those focusing on emergency food or emergency shelter. A few requests for service are handled by recorded messages because the information is easily conveyed. For example, individuals calling with problems about their utility bills are automatically provided with the telephone number of representatives of the appropriate utility company. Interestingly, however, the major innovation at INFO LINE in recent years has been a move toward the establishment of neighborhood centers, which can provide a walk-in service and also locate small, storefront service providers within a neighborhood that would not otherwise be visible to their centralized information and referral database. In Los Angeles County alone, INFO LINE maintains information on over 5,000 service providers. Its first two neighborhood centers were established in a Korean-American community and a distressed African-American community of South Central Los Angeles.

A number of jurisdictions have experimented with electronic bulletin boards and other online systems for offering dial-up on-line access to databases. From 1987 to 1988, a committee of the California State Legislature developed The Capitol Connection, a dial-in bulletin board of use to individuals wishing to keep up with the committee and the legislature's agenda. Libraries have been among the most likely agencies to develop on-line services, allowing computer users to access the library's card catalogue from their home or office. The public library in Pasadena, California, expanded on this concept to develop a database of civic organizations, city commissions, and social service agencies as well as a community calendar listing everything from performing arts programs to garden club meetings; they call the system PALS, the Public Access Library System (Guthrie, 1991). The library of the City of Glendale, California, plans to launch one of the most technologically state-of-the-art public access computer systems, which they call the Local Information Exchange (LNX).

A few other cities are using computer bulletin boards to provide citizen access to information. The city of Santa Monica, California has led the way with its Public Electronic Network (PEN). PEN contains information on community events, the council's agenda, as well as other information on city departments, officials, and services. It is accessible from personal computers or any more than two dozen public terminals distributed throughout the city (Dutton & Guthrie, 1991). The New York City Board of Education has also developed a computer bulletin board called NYCENET, used by both educators and students.

A number of government agencies are considering the provision of on-line access to specialized databases, both as a means for improving services as well as a way to generate new revenues. For example, some agencies, such as in Arlington County, have considered providing on-line access to land records. They believe some title companies might be willing to pay for the convenience of accessing these records from the company's offices. If land records are automated for other reasons, this kind of service is increasingly practical at only a marginal cost. Hampton, Virginia has provided 25 incoming lines to their computer, where people who subscribe can access information such as the sales history of property, the owner, city tax assessments, school assignments, and permits. Geographically based information systems that integrate census, housing and other information are another specialized database that is potentially of value to telemarketing firms.

On-line access to records, increasingly stored on optical disc imaging systems and integrated with facsimile equipment, provides an attractive alternative for private firms to use in accessing records and a promising new revenue source for the public sector. Information can be retrieved for on-screen viewing or directly "faxed" to the user.

"Interpersonal communications" comprises the fourth category of examples. Government officials and personnel have long communicated with the public in person, over-the-counter, by mail, and by telephone. Increasingly, they are using voice mail, facsimile, electronic mail, and computer conferencing systems. Information and referral agencies rely very heavily on interpersonal communication between the public and information specialists. The information specialists often provide counseling and emotional support as well as information to their clients. One important innovation in information systems is designed to support the communication between client and information specialist by providing the information specialist with an expert system to assist in the identification of appropriate agencies for referral. As being implemented by INFO LINE in Los Angeles, the information specialist will input data on the client's location, age, gender, any handicaps, and needs. The system will identify the most proximate and relevant agencies for referral. Eventually, this system will replace a heavily annotated manual file of information on 5,000-plus agencies that sits in front of the information specialist. It should speed and improve the delivery of information to clients and overcome some of the inherent limitations of a service that is too heavily dependent on the memory of specialists.

Voice mail is viewed as a means to reduce the staff time required to handle routine telephone calls. It is also being used to enhance rather than replace interpersonal communication, for example, by supporting parent-teacher communication, allowing the teachers to leave a message at the end of each day and providing the working parents with a means to hear the teacher's message or leave a message for the teacher. Other state and local agencies are using voice mail for answering inquiries about teacher certification, filing insurance claims, and inquiring about the status of claims.

Cities are also using facsimile machines to receive and, more recently, distribute forms. Cities can send and receive applications for permits via facsimile. Optical disc storage systems can be used in conjunction with facsimile machines to distribute records identified through an online search.

A potentially more dramatic departure is the use of electronic mail. This technology has been adopted since the 1970s by government agencies for internal, organizational communications, but a few governments have begun to apply this technology as a means to support dialogue with the public. The City of Santa Monica's Public Electronic Network (PEN) stands out in this area. PEN allows citizens to send electronic messages to city hall, council members, any city department, and the library's reference desk. City staff can then reply to the request electronically. City departments received more than 5,000 messages from the public in the first 2 years of PEN's operation. Sixty percent of the individuals described their messages to city hall as inquiries. About one fourth sent either comments, complaints, or other kinds of messages to city officials (Dutton, Wyer, & O'Connell, 1993).

Some city department heads and supervisors in Santa Monica were reluctant to implement electronic mail between citizens and city hall for fear it would open a floodgate for complaints. In PEN's case, it does seem to have increased the number of complaints and requests coming to the city. Therefore, this facility is perceived to have increased the workload on city staff but also the responsiveness of the city to the general public (Dutton et al. 1993). PEN is also innovative in its provision of facilities for citizens on the system to send electronic messages to one another as well as to the city.

This augments PEN's use for electronic meetings and conferencing about public affairs. In Santa Monica and elsewhere, electronic mail and conferencing software are being used to support public forums on policy issues. In contrast to electronic voting and polling systems proposed in the 1970s, these forums are designed to foster dialogue to set agenda and explore options rather than facilitate voting and polling from the home (Dutton et al. 1991). While Santa Monica's PEN system is the only municipally funded project that has extensively developed a conferencing capability, political forums have been developed on Cleveland FreeNet, a nonprofit operation in Cleveland, Ohio, and by Berkeley's Community Memory project. The New York City Board of Education's NYCENET System also features discussions (Dutton et al. 1991).

Finally, in the interpersonal category, government agencies are using audio and video teleconferencing in creative, but controversial ways to reduce travel and increase communication with remote locations. One instance is in the criminal justice area, where a number of jurisdictions are trying to reduce their costs for transporting inmates by providing a two-way video link between their jails and courts. People arrested for misdemeanors, and in some jurisdictions for felonies, can be arraigned and receive a bond review from the jail to which they are taken after their arrest.

There are a variety of other applications of electronic media that are increasingly moving citizens in more direct, albeit mediated, interaction with government agencies. Advances in computer-assisted dialing and interviewing systems are being used by public opinion polling firms and could be used by governments in similar ways to gather information about the public. The entertainment media are experimenting with new versions of interactive television that utilize over-the-air broadcasting, in home terminals, and the telephone network to poll viewers as they play along with televised game shows. The same technology could be used by government agencies to interact with the audiences of public or governmental affairs programming, much as they did in

Columbus, Ohio, when the QUBE system, developed by Warner-Amex in the early 1980s, was still in operation (Davidge, 1987). Such systems could and have resurrected interest in voting and polling from the home, even if it is problematic that they will ever diffuse to a larger proportion of the public. Also, the ordinary push-button telephone could be used increasingly for polling citizens on a variety of issues—either alone or in conjunction with televised speeches or debates. The introduction of screen-phones, initially being designed to facilitate voice processing by presenting options over a liquid crystal display device mounted on the phone, will undoubtedly encourage such applications.

There are also examples of the use of video and other electronic systems for monitoring the public, even though they are another application that will increasingly bring the public in more direct linkages with governments. For example, transportation agencies plan to employ video more routinely as a means to monitor traffic conditions in real time. Caltrans, the California Department of Transportation, is developing an automatic vehicle identification system (AVI), much like that developed in Singapore, to automatically charge for tolls. Singapore uses its system to control traffic by charging additional fees for the use of streets within the central business district during peak rush hours.

NEEDS IN THE FORM OF ACCESS TO INFORMATION

In theory, the interests and needs of the public should guide technological change. In practice, it is difficult to identify needs far enough in advance and in clear enough terms to shape the design and development of technology. However, our survey of innovations in electronic service delivery along with more focused research in the inner city surfaced a variety of public needs and interests that are embedded within the design and development of innovative electronic services (Dutton, 1992b; 1992c). We found quite genuine needs among the public for information services, whether or not they are driving the development of electronic services. Broadly, they can be classified into two general types of needs: access to information and access to information technology, the new medium for obtaining information and communication services. We address need for information access first.

In an era of so-called information overload, few managers or professionals can imagine situations in which there is truly a lack of essential information. Ironically, that is precisely the case, particularly among the less well-to-do.

For example, while it is well known that many people go without food or shelter due a lack of a sufficient number of food programs and shelters, it is less well known that some individuals simply do not get needed food and shelter for the lack of information on its availability. INFO LINE, a nonprofit agency, offers public information and referral services over the telephone to anyone calling within the county of Los Angeles. They receive from 220,000 to 230,000 calls per year and have been processing about the same number of calls over the last 3 years, despite deepening problems with unemployment, homeless, and public services within the county and city of Los Angeles. One of their managers argues, convincingly, that the number of calls they can process is primarily a function of staffing. If they doubled staff, they might be able to double the number of calls they process.

Currently, there is an average wait of 7 or more minutes per call before an information specialist can speak with the client. They have no way to track the number of people who hang up before their call is answered, but they realize it is a high number. In 1991, out of a total of 204,351 tabulated calls, 12% were for emergency food, 9% for emergency shelter, 5% for utility bills, and the remaining reflected 482 different categories. Given that the two most frequent calls to the agency are for emergency food and emergency shelter, a failure to get information is potentially depriving individuals of a meal, a place to sleep, or other basic needs.

Other examples of information bottlenecks exist. One

that potentially affects every citizen is the processing of "911" calls in major cities, including Los Angeles. According to some authorities, it takes up to 10 minutes to answer some 911 calls in the city of Los Angeles. One survey of over-the-telephone operators in both the private and public sector found that one of their most constant frustrations is apologizing to callers who are continually upset because they have not been able to get through on the phone (Fountain, Kaboolian, & Kelman, 1992).

A parallel can be found in the area of broadcasting, where there are also needs in the midst of abundance. At the same time that many communities can get access to dozens of channels of entertainment television over cable and thousands of videocassette titles, many linguistic minorities in the United States are poorly served by the mainstream news and entertainment media.

For example, in the disturbances in Los Angeles following the verdicts in the Rodney King beating, the Korean-American community was very dependent on local Korean broadcasting, primarily radio, and local Korean newspapers to keep their community informed about the course of events. In the hours after the disturbances broke out, Korean-Americans quickly concluded that the mainstream media did not know what was going on. There was a clear sense within the Korean-American community that Korean-American shops were being targeted by looters, but there was no mention of this possibility on the mainstream media, which set the agenda for the police and politicians within Los Angeles.

Working with Korean broadcasting, leaders within the Korean-American community were able to marshal enough evidence to demonstrate their case to the mainstream media. Eventually, coverage by the mainstream media captured the attention of politicians and police agencies, who then directed more resources to help Korean-American and other small shop owners.

This case is but one example of an enduring problem of encouraging minority-oriented programming by commercial broadcasters. While critics have long noted the need for minority-oriented programming for television, the equivalent need for minority-oriented software for computer systems is only emerging on the public agenda. Software for a personal computer is quite analogous to videocassettes for a television equipped with a videocassette recorder. Similarly, computer software that is an increasingly significant commercial product is as likely to be oriented to the mainstream culture as is a television program.

Cultural barriers are tied in a variety of ways to educational barriers. Education has long been viewed as a barrier to access for minorities. Skills in using many technologies are supported by traditional education, specifically reading, writing, and arithmetic, which offer a gateway to functional literacy within a variety of areas. Most computer software and its associated documentation assume at least a high school education and knowledge of the English language as well as a background within the mainstream culture.

These cultural and educational barriers create a general need for bilingual staff and multilingual services tailored to the diversity of languages and cultures that compose our society. In the case of California, more than one third of residents are Hispanic or Asian, and 6.5 million individuals in California are considered "language dependent," which means they depend on a language other than English. The Los Angeles County Library offers services to more than 45 different language communities (Dutton, 1992c). In diverse, multicultural communities like Los Angeles, experiments with bilingual information kiosks, for example, have been judged successful by public agencies, such as the Los Angeles County Library. The L.A. Project, which provided information on about 200 social services over multimedia personal computer kiosks, offered three language options—English, Spanish, and Korean—with some success. Commercial services, not necessarily developed with distressed populations in mind, such as AT&T's translation services, have often been of value to minority residents and agencies. Information specialists at

INFO LINE, Los Angeles County's information and referral agency, provide services in English, Spanish, and several other languages, but set up conference calls through AT&T's translation service when they encounter a language they are not prepared to serve.

Multilingual services require more than simply offering services in more than one language. Every service needs to be viewed from the perspective of a non-English-speaking client. For example, INFO LINE found that many Spanish-speaking clients would hang up if they heard a message recorded in English before they could hear the message repeated in Spanish. INFO LINE experimented with some success in switching the order of presentation, offering the Spanish instructions first. Similarly, their staff found that a flyer inviting individuals to call "213-OPEN" simply confused their Spanish language clients.

Political jurisdictions create another barrier to public access and, therefore, a need that might be addressed by electronic service delivery. Generally, departmental and governmental jurisdictions often fail to conform well to the information needs of the public. Political jurisdictions need to be made increasingly transparent to the public, who want information from wherever it lies. Computer and telecommunication networks could bridge political jurisdictions and permit individuals to access information from any jurisdiction. Computer software is already available that will allow the users of many university computing networks to locate and retrieve information by subject rather than by the particular university computer on which it is stored. Analogous software could be used to create a network of nonprofit, local, state, and federal agency computer networks that would technologically cross jurisdictional barriers to gaining access to information or services.

In experimenting with a touch-screen information kiosk, staff of the Hall of Administration of the County of Los Angeles saw immediate disadvantages in the fragmentation of services among multiple departments and jurisdictions.

The public did not want to move from kiosk to kiosk to find information or services. To the county, the fragmentation of information across county departments and local government jurisdictions was a major constraint on the success of their experiment. The provision of information simply from the Hall of Administration did not provide a sufficient scope or range of services to provide a significantly greater advantage to using a kiosk versus waiting in line at the same facility. To create a successful information kiosk, it seemed important to provide an ability to cut through the boundaries of departments and jurisdictions so that the public could find the information they sought whether it was collected and maintained by the State Department of Motor Vehicles, the County Hall of Administration, or the City Attorney's Office. Just as the banking industry has permitted access from nearly any automated teller to nearly any bank or credit card company, despite the firm or location, so should a public kiosk provide access that overcomes the institutional fragmentation of the public sector.

This jurisdictional issue extends beyond the public sector. For example, a month after the disturbances of late April 1992 in Los Angeles, I tried to call the Federal Emergency Management Agency (FEMA) from my office in the central city of Los Angeles. The telephone operator first told me that there was no listing for this agency. I eventually discovered that their office was nearby, in Pasadena, but within a different area code and, therefore, off the radar screen of the local telephone operator.

One of the most formidable barriers to creating truly valuable information systems for the public is cost. Anyone who has even set up an electronic directory of addresses and phone numbers quickly learns that the development of a complete and up-to-date database is a major investment of time and energy that must be constantly attended. North Communications was able to set up six touch-screen information kiosks around South Central within 12 days after North launched their initiative; the kiosks contained information on about 200 services. Among the lessons learned by North Communications was that "getting enough information into the system quickly enough" was one major requirement. They also discovered that the data was highly volatile and that they needed to make frequent updates.

North Communications was not alone. According to one of the top managers of a Los Angeles information and referral agency, in the aftermath of the disturbances in Los Angeles, "many agencies and elected officials felt that they had to establish information lines, but few of the lines they started had much value" (Wallrich, personal correspondence, 1992). As he put it: "A needless proliferation of information services is not good at any time, and especially should be discouraged during a crisis, when inaccurate information can be very harmful."

The size and complexity of this undertaking is illustrated by INFO LINE, which collects and maintains information on over 5,000 social service agencies, all within the single county of Los Angeles. Even so, the manual files on case workers' desks are filled with annotations and corrections made on the basis of their day-to-day telephone contacts with agencies. Ironically, anyone can start a database or information service, but developing a timely, comprehensive, and accurate database is a far more significant undertaking than is often understood by would-be information providers.

Finally, the harried pace of working families often makes it difficult for individuals to attend meetings during the day or evening hours, which lessens the adequacy of traditional approaches to informing the public. Our survey of innovative projects highlighted a number of efforts aimed at addressing a perceived decline in their ability to reach "live audiences." Neighborhood television projects, such as that in West Hartford, Connecticut, are based in part on a sense that it is increasingly difficult to get individuals to attend meetings, hearings, and other live events. Televised events are not meant to substitute for attendance, but to provide some opportunity for residents to be informed when they would otherwise not be able to see the event at all. Likewise, a sizable number of PEN users are active in local political discussions by virtue of their access to PEN, when otherwise they would not have been involved in public affairs because their family or work requirements would not permit them to attend regularly scheduled meetings.

Indirect evidence that electronic network services might diminish some barriers to communication between the public and institutions emerged from our research on PEN. We surveyed 72 personnel in the city of Santa Monica about their perceptions of the impacts of PEN on their work and the responsiveness of the city to the general public. They saw increased impacts of service requests (83%), variety of work (55%), time pressures (53%), complaints (45%), and staffing needs (38%). The general pattern of their responses provides a convincing case that the city's personnel believe that they are getting more requests for service, face more time pressure, and have become more responsive to the public in Santa Monica as a result of the PEN system.

Many cling to early visions of a mass market for the new electronic information services within the near future. However, most new media businesses have focused on niche markets. For example, while newspapers and other commercial service providers might eventually succeed in offering information to a sizable proportion of the public, most have focused on electronic services that complement the newspaper, which they continue to view as central to their future. In the fall of 1992, *The Palm Beach Post*, owned by Cox Enterprises, began offering some weather, sports, and other information over an "easy-to-remember" three-digit telephone number (Guy, *USA Today*, October 28, 1992). If such services can be offered at an incremental cost to newspapers and their subscribers, they might well become profitable even if utilized by a small proportion of the public.

As already discussed, most of the successful information and communication projects in the public and private sector are oriented to serving information needs that are far more specialized, complex, and personal than often suggested by discussions of electronic publishing or for that matter the public sector equivalent of the automatic teller machine (ATM). Contrary to the problems facing the banks and financial institutions, the problems of creating a public information service are not those of designing 20 screens of information that will be accessed billions of times. Quite the contrary. Public information needs are so personal that it is more likely to be a problem of designing thousands of screens of information that are uniquely tailored to a specific individual with a unique set of needs at a particular time and place. Some of the most successful innovations are responsive to non routine and specific questions or needs that are not shared by a large proportion of the public. Examples abound.

On Santa Monica's PEN system, there are literally hundreds of on-going discussions underway, many of which address the particular interests of only a handful of individuals. The French Minitel terminals provide access to over 15,000 different services. SeniorNet is focused on the needs of elders, particularly health care information (Arlen, 1991). Community Link, a joint effort of The Center for Community Change and Apple Computer, Inc., seeks to serve organizations that are focused on the economic development of low and moderate-income communities. The network provides information on such matters as trust funds, financing, Community Development Block Grants, and subsidized housing, as well as a channel (electronic mail) for community advocates to communicate with one another about their common problems. HandsNet, another nonprofit information and communications network, is also specialized, focusing on serving agencies providing human services.

The specialized information needs of individuals are also evident in the activities of information and referral agencies. INFO LINE keeps records of the requests for information it receives. Over a 1-year period, the requests from over 200,000 calls fell into nearly 500 categories. The only categories that might be called routine were the three most frequent requests, which were for emergency food, emergency shelter, and assistance with restoring gas, electric, or other utilities. An index of public information needs developed by the INFO LINE staff took years to develop and fills a 3-inch-thick directory that looks like the telephone book for a major metropolitan area.

The proliferation of audiotext "900" numbers is another reflection of the diversity of individuals' information needs. Specialized services, such as *USA Today*'s "The College Info-Line," that can be of value to only a small proportion of the public but on a national or regional scale, seem especially well suited for this type of specialized service.

NEEDS IN THE FORM OF ACCESS TO TECHNOLOGY

Equity is one of the most central issues raised by the critics of electronic service delivery. Many fear that electronic access will widen disparities between the information rich and the information poor, since the less well-to-do are unlikely to have the income, slack resources, or exposure to computing at work that will facilitate their access to and use of information technology as a medium for communication. In brief, access to technology is itself a need issue.

In Santa Monica, this issue was raised early in consideration of the launching of PEN and led the city to establish a number of public computer terminals from which individuals could access the PEN system. Even though nearly one third of Santa Monica households have access to a personal computer, almost one fifth (19%) of all accesses to the PEN system have been from one of the public terminals available at libraries and other public sites within the city (Dutton et al. 1993). The Department of Information Services has even had to develop rules, based on those established for sharing tennis courts, to govern how long an individual can stay at a terminal when others are waiting to use it.

In September 1992, months after the disturbances in Los Angeles, I organized a community workshop on electronic service delivery and the inner city. I expected that the workshop would focus on the risks to minority communities and cultures of innovations in electronic services. Innovations in electronic service delivery pose a number of threats to the quality of public service (e.g., equity of access, the protection of personal privacy, or the retention, archiving, or accuracy of records). Overall, however, it was remarkable how few problems, threats, or risks became the focus of discussion. Possibly due to the influence of early presentations or the bias of most participants, the thrust of discussion focused on ways to bring technology into the inner city rather than concerns with technology eroding the quality of existing services.

In many distressed areas of the United States, particularly areas of the inner city, there is a very basic need to expose children and their parents to the existence and utility of information technology. The technology is simply invisible to many families. One of the primary reasons that households adopt a personal computer is to learn more about the technology per se (Dutton, Rogers, & Jun, 1987). Mere exposure to information technology is informative, irrespective of the use of the technology as a medium for accessing other information (Dutton, 1992c).

However, there are a number of identifiable needs for access to information technology within distressed areas. Interviews with public and voluntary organizations in distressed areas surfaced a need for systems to support scheduling and coordination among a variety of specialized agencies, such as central city schools. Currently, they lack good channels for communicating with one another on a day-to-day basis (Dutton, 1992c). One principal of an inner city magnet school in Los Angeles argued that her school was electronically isolated in the sense that she lacked adequate facilities to schedule and coordinate activities with other nearby and suburban schools, resulting in a failure to take advantage of events, space, and other resources that could be more efficiently shared. The plain old telephone network was not sufficient to meet her needs and, therefore, she perceived a need for an electronic system that could support more asynchronous and efficient scheduling and coordination, such as an electronic bulletin board.

Access to technology is also perceived to be a need of minority business firms, which often lag behind more established firms in their use of information technology. This gap places the minority firm at a disadvantage, such as in competition to become primary or major subcontractors on projects. Reasons why they lack the technology are in part educational but also financial—they do not have adequate funds to purchase state-of-the-practice computers and telecommunication equipment. If they do purchase the equipment, they are often placed in a position of buying dated models and lower-end systems that are more affordable. They must wait for systems to trickle down to their price ranges for them to be affordable. However, that situation leaves them technologically behind their major business competitors, which are using state-of-the-practice systems and who can take advantage of breakthrough technologies when they arrive.

Technology is not simply equipment, but also the know-how and expertise involved in using the technology. In this respect, there is a clear need for greater access to expertise and technical assistance in computing and telecommunications, particularly on the part of many voluntary, not-for-profit, and minority business firms. They often lack the technical consultants within their ranks, which are widely available to established business firms, who can advise and assist personnel with computing and telecommunication problems. Even if an individual can use a personal computer with some ease, that person might be brought to a standstill by a problem logging onto a online network, setting up their "autoexec" file, or any other of a countless number of minor challenges. One not-for-profit agency, CompuMentor, has been organized primarily to provide on-site assistance with computing to voluntary and not-for-profit enterprises, but the gap remains.

"COMMUNICATION" AS A CITIZEN INTEREST

Early visions of the public information utility were based on the assumption that the public is interested in information, but this assumption might well be a misleading guide to the development of services. In fact, many assume that the public is composed largely of avid information seekers. However, the experiments with electronic services we surveyed tend to reinforce an observation made by many involved in the early market trials of videotex services, which is that the public might be more interested in specialized services and communication than in information per se. As Richard Hooper (1985, p. 190) put it in discussing the lessons learned in the marketing of Prestel, Britain's innovative videotex service: "Prestel was invented and launched on the assumption that there is a large and ready market for electronic information retrieval—retrieving pages of information stored in computers. Reality turned out differently."

This point is striking from our research on the Santa Monica PEN system. PEN provides a menu of services, which includes several options for electronic access to static information about the city, the council, and community events—a read-only menu to bulletin board services. However, PEN's "Mailroom" provides the capability for electronic mail. From the outset of the PEN system, conferences generated more interest, as gauged by the number of accesses to this facility, than did the bulletin boards. Moreover, there is a trend toward an increasing focus on conferences and electronic mail as opposed to simply retrieving information stored online. This might in part be due to the relatively static nature of information about the city and community events as opposed to the content of mail and conferences, which can be quite dynamic, but it may also reflect the public needs and gratifications served by communicating with other individuals as opposed to simply obtaining information.

This phenomenon is not new. The ARPANET was originally developed to support remote access to computing services, but evolved instead to become primarily a medium for interpersonal communications. SeniorNet, HandsNet, Community Link, Cleveland's FreeNet and other computer-based information services for the public have made electronic mail and conferencing a central aspect of their operations. In fact, as we found in observing a number of agencies that are expressly focused on providing "information" to the public, the boundary between information and communication can be quite problematic. Even agencies that are explicitly devoted to the provision of information services can find themselves offering a communication service. For example, information and referral agencies, such as INFO LINE, are called by individuals needing particular information, whether it be food, shelter, counseling, financial assistance, or other services. Even the most casual observer of the information and referral function could not help but see that these calls normally generate an extended conversation about the nature and history of one's situation and a set of recommendations regarding the options and problems facing the individual. The information specialists provide counseling and emotional support of another human being—not just information.

The distinction between communication versus information services is not simply a semantic issue because it has significant implications for policy and practice. Instead of focusing attention on providing access to specific information content, which is a major theme of many scholars interested in public access, we should focus more attention on access to information technology as an emerging medium for communication. It is access to the technology as a medium for communication versus specific informational content that is paramount. Similarly, we should view the public less as an audience or even as users of information services, but as providers of content and services. This realization could also change the way we look at the information service agencies themselves, since they might be better conceived of as networkers or communication firms than as content providers or publishers.

Of course, the reality is that both perspectives are true with the new electronic services. The public is both audience and provider; the public needs access to information as well as to information technology; and information providers are publishers but also carriers or transmitters of information provided by others. The point is that the significance of the communication functions served by the new media is too easily and inappropriately marginalized in discussions of access to public "information" unless information is defined in its broadest sense.

SOME GENERALIZATIONS

The primary rationale behind promotion of the public information utility in the 1960s, like that behind videotex in the 1980s, was the public's interest in getting information. The market failure of interactive cable television and, later, of commercial offerings of videotex services undermined this rationale to a great extent but not entirely. Many developers of information systems continue to believe that a substantial proportion of the public is actively seeking information and will use new information systems if the right information is packaged in appropriate ways. This is one line of reasoning behind the growing interest in multimedia personal computers and kiosks as well as new telephone services, such as the screen phone, for the delivery of electronic information services.

However, the public and not-for-profit sectors are driven also by a variety of other rationales to develop mechanisms for electronic service delivery. Some distressed segments of the public lack basic information of relevance to their health and safety. Among politically active citizenry, such as in Santa Monica, a growing number expect more direct and convenient ways to participate in public affairs and communicate with public officials and agencies. Public as well as not-for-profit agencies share an interest in reducing the costs of providing routine information to the public and stand to benefit from systems that will support communication within communities in which traditional forms of interpersonal communication are perceived increasingly to be insufficient. The growing prominence of ethnolinguistic communities demands that public agencies be more responsive to the cultural diversity of their clientele.

The success of the public and not-for-profit sector in using electronic media will create a need to address issues over the equity of service provision across jurisdictions and socioeconomic groups. Equity considerations sometimes moderate interest in cutting edge technologies, since their expense is likely to reinforce socioeconomic disparities, but these same equity concerns can support the extension of more universal access to electronic communication and information technologies that are already in use (Williams & Hadden, 1993).

Despite a number of genuine needs for extending electronic information services, these experiments face formidable problems. One is that the information needs of the public are complex, multifaceted, and unique to individuals, making it difficult and costly to adapt systems, such as an information kiosk, that are more easily used to provide easily programmed responses to routine inquiries, such as a bank balance. Another is that information resources are often so fragmented across jurisdictions and organizations that they do not match well with the needs of individuals. Information that is relevant to these needs is exceedingly difficult and costly to create and maintain in a way that is comprehensive, accurate, and up-to-date, resulting in only a few truly valuable information systems that serve the public interest.

Electronic service delivery is not a quick technological fix to fundamental problems of citizen access, which often extend from the growing diversity of urban centers, the financial crises facing American governments, and the basic needs of the poor and unemployed. However, failure to develop the infrastructures and applications to provide all sorts of electronic services to the public might well worsen these problems within an information society in which all kinds of services are increasingly being mediated by broadcasting, telecommunication, and other electronic media. Information technology can be used to facilitate citizen access, to make agencies more responsive to the public, and to better meet the needs of many specialized publics for information about health and social services. Electronic information services are raising problems of their own, including concerns over the equity of electronic service provision and the rules governing the privacy, confidentiality, and freedom of expression over electronic media. If a patchwork of isolated electronic service delivery systems emerges, they might well exacerbate rather than diminish inequalities in access to information resources, not just to the poor, but also to any member of the public who faces educational, language, or physical handicaps that create a barrier to accessing information or services.

The appropriate national policy is more complex than suggested by much of the discussion of "a national information highway." The federal government might undermine the development of electronic service delivery by promoting any one medium for reaching the public with information perceived to be in the general public interest. If the emerging trends discussed above are any guide to the future, federal strategies for supporting the public's information and communication needs should focus on facilitating the migration of all sorts of specialized information and services to all kinds of electronic media, including broadcasting and telecommunications as well as newer information technologies like voice processing, computer bulletin boards, and multimedia kiosks. Federal, state, and local agencies need to support national initiatives to coordinate the diffusion of electronic networks and services throughout the United States.

At the same time, the public and not-for-profit providers of information services must insure that the public has the capability to access electronic information; this access will come through the provision of public facilities and support to groups that will be electronically disadvantaged, if not disenfranchised, without vehicles and "on ramps" to the many information highways criss-crossing the United States.

There is also a federal role in tying together the growing multiplicity of electronic networks and services created at the federal, state, and local level. It can do this by setting standards, interlinking networks, and providing an indexing function so that the physical location of information or agencies is increasingly irrelevant to the citizen needing access to a particular service. In these ways, innovations in the provision of electronic services by public and not-for-profit agencies at the local level can be complemented by the development of regional, state, and federal initiatives to construct a seamless national network of networks for the public as opposed to an electronic maze of stand-alone applications.

One technically rational but dubious assumption of the age of electronic information networks is that it no longer makes any difference where information might be physically housed. If you work or sit at a computer, so this argument goes, modern telecommunications enable you to work as easily from the opposite coast as from the same office in which the computer is located.

One theme emerging from a variety of relatively successful public and not-for-profit information systems is that systems benefit greatly from a psychological sense of community ownership. SeniorNet's success is in part a function of its perceived ownership and control by seniors. Users feel a part of a community, rather than simply the market of a new media entrepreneur (Arlen, 1991). PEN users are PENners and residents of Santa Monica—literally members of a community. Community leaders in the inner city of Los Angeles do not want their communities viewed as a new market for information technology; they want to gain a sense of ownership and control over information technology that they design to serve their own community's needs (Dutton, 1992a).

It might well be that the development of any national information system must always balance efforts to achieve economies of scale and scope with efforts to develop and maintain a sense of community and ownership. Those who are expected to benefit from the use of particular services need to truly feel that they are being served by a system that is controlled by their community. Their community might be defined by categories of needs and interests, such as senior citizens, or by geography, as in the case of Santa Monica residents.

Political jurisdictions, along with not-for-profit and other quasi-governmental agencies, can become valuable providers of electronic services that the public can access from their homes and public facilities over a variety of media. However, their efforts will be limited if not diminished if Federal policies do not create incentives and mechanisms to provide the on ramps, off ramps, and highways to make it more universal and equitable to the American public. But it will not be neatly integrated on a single, national electronic network. Instead, it will evolve bit by bit as specialized systems developed by early innovators, such as those discussed in this chapter, diffuse and become linked into larger regional and statewide systems of public and not-for-profit networks that support broadcasting, transactions, access to public information and records, and interpersonal communications over broadcast, cable, and telecommunication networks. The wired nation will be a complex and decentralized mosaic of specialized networks rather than a single, federally planned and integrated information highway.

ACKNOWLEDGMENTS

This chapter is adapted from a longer report prepared for The Freedom Forum Media Studies Center at Columbia University. The author wishes to thank K. Kendall Guthrie and Cherilyn Parsons for their comments, and acknowledges the Office of Technology Assessment of the U.S. Congress for underwriting parts of his research.

Chapter 6

CITIZEN ACCESS, INVOLVEMENT, AND FREEDOM OF EXPRESSION IN AN ELECTRONIC ENVIRONMENT

BY JOHN V. PAVLIK

Absent from much of the debate about the emerging information highway are the implications of electronic communications on the everyday citizen. How will all citizens gain access to the information highway? Will the interests of all citizens help shape the design of the new communications infrastructure? How will the guarantees of the First Amendment be protected in an electronic communication environment? This chapter directly examines the views of seven leading communications scholars and analysts who responded to these questions at the October 27 national conference: Media, Democracy and The Information Highway. Julius Barnathan, former senior vice president for Technology and Strategic Planning at Capital Cities/ABC, Inc., pioneered the development of closed-captioning for the hearing impaired. Stuart Brotman, a communications lawyer and senior fellow at the Annenberg Washington Program, Northwestern University, has written extensively on new technology and people with disabilities. Jannette Dates, a professor in the School of Communications, Howard University, is the co-author of Split Image: African Americans in the Mass Media. *William H. Dutton is national director of the U.K. Programme on Information and Communication Technologies (PICT) at Brunel, The University of West London. Currently on leave as a professor of the Annenberg School for Communication, at the University of Southern California, Dutton has written widely on communication technology, including chapter 5 in this volume. K. Kendall Guthrie is director of Telecommunications Planning in the Department of*

Telecommunications and Energy for the City of New York. Monroe Price is a professor in the Cardozo Law School and has written extensively on First Amendment issues and new technology. Frank Blethen is the publisher of the Seattle Times *and has pioneered the use of electronic services in news delivery.*

BUILDING BLOCKS FOR A NATIONAL INFORMATION SERVICE

Perhaps the best place to start any discussion of citizen access to a national information service is with what communications lawyer Stuart Brotman called the bible of U.S. communications law and policy: The Communications Act of 1934. Section One, Title One, of the Communications Act is a mandate by Congress for universal service. Although the term *universal service* does not appear in the Act, the intent of the Act is interpreted and often referred to as such. The Act states that there shall be made available, as far as possible, to all people of the United States, a rapid, efficient, nationwide, and worldwide wire and radio communication service with adequate facilities at reasonable charges. In other words, for more than 50 years the United States has had a national communication policy aimed at universal service. One of the major challenges in this decade and beyond is defining universal service. As new technologies are developed, we have to give new meaning to the words of this bible. Specialized needs and new needs will arise for communications technologies that may not quite fit into the original notions of universal service.

One of the most important building blocks of a national information service centers around the issue of access to the information highway, the wired network that will be the conduit for information. In a nation dependent on an informed citizenry, all segments of society must have equitable access to the information that can help improve their circumstances.

This means the information highway much reach rural areas and inner cities, and serve those with housing, clothing, and health-care needs.

ASSESSING NEEDS ON THE INFORMATION HIGHWAY

Although it is extremely difficult to determine people's information needs, an examination of information systems and communication technologies that are being initiated by American local and state governments, as well as nonprofit agencies, reveals two clear conclusions. One is that the public information utility, or the national information service in some emergent form, is already here. It is no longer a futuristic pipe dream or utopian vision, according to telecommunications scholar William H. Dutton. Hundreds of on-line commercial and public information services are now available throughout the world, with a variety of service providers, including newspaper companies, telephone companies, and financial institutions. The services are delivered over a variety of telecommunications media, including fiber optics, traditional telephone lines, and in the near future, via satellite. Major national and regional consumer on-line information services now reach nearly 3,480,000 households in the United States (Arlen, 1993a).

These emergent services already are changing the relationship between citizens and government. Examples include the FreeNet community-oriented service now available in some two dozen cities around the world, including the United States, Canada, and Germany and the Public Electronic Network in Santa Monica, California. These and other services are free to all users and although overall usage is low (i.e., these services have a combined total usage of about 70,000 people each week), the usage rate for these municipal services is rapidly growing. FreeNet is the largest of these public services providing information about community services and government agencies free to all users. Nearly 50

FreeNet services will be operational by the end of 1994 (Arlen, 1993b).

Rather than looking at a romantic past or utopian future, we need to study what is being done presently with such services. We already have an idea of what the public wants and which services are successful. The issue is how to influence the development and design of these systems so that they benefit society. The electronic kiosk system is an example. A kiosk provides information over a public terminal typically using a touch-sensitive screen. Informational kiosks are seen in a variety of locations, such as shopping malls, grocery stalls, government offices, and public libraries. When a welfare client fills out an application at a kiosk, it changes the way the information is collected and alters the way people interact with government. Significantly, this accessible, easy-to-use tool could become a useful model for more sophisticated tasks, like paying taxes, renewing licenses.

A second conclusion is that no single model completely describes what is being done today, or what will work in the future. Videotex is not the model for what is being done. Nor is the personal computer, the information kiosk, the smart telephone, the screen phone, the facsimile machine, or even the electronic town hall. Each of these technologies is used for different types of services. Some are effective for broadcasting information, others for transactions, information retrieval or interpersonal communications. It is an extremely diverse array of services. However, four aspects of needs are evident in these systems.

First, the systems are extremely specialized. Because the information needs of the public are very complex, personal and specialized, the information systems must be so as well. For example, INFO LINE, an information and referral service in Los Angeles, overwhelmingly receives only three common questions. The most common involves emergency food. The second is for emergency shelter, and the third is when the caller's utilities have been cut off. Beyond these three questions, there are no commonalties and about 500 categories of questions, each receiving about five or six calls during the year. So providing universal service is not a question of designing 20 screens of information that will be utilized a million times. It is also an issue of providing answers to very particular questions for a handful of people in Los Angeles county. In the Santa Monica system, there are hundreds of discussions going on simultaneously; there is not a single town hall meeting. Rather, there are many meetings, with few talking to few. People discuss very specialized topics and have specialized interests. Among the existing information services, from Minitel to SeniorNet to Prodigy, all the successful services are very specialized, serving particular networks and communities.

A second dimension is that people are interested in sending and receiving messages at least as much as in retrieving information content. Although these two features often overlap, it is important to note the distinction. If you look at the evolution of Bitnet and Internet, people by and large use these systems for sending messages, not remote computing. Among users of the Santa Monica system, information retrieval is increasingly a minor part. Even in Knight-Ridder's short-lived Viewtron videotex system in the 1970s, one of its most successful applications was electronic messaging. Person to person messaging, electronic mail and conferencing are the dominant uses of the Santa Monica system. This is true on almost all of the other computer bulletin board systems. Not only do the people want to talk to each other, they want their own network, topics and conferences. In short, they want their own system. Thus, it appears that access strictly to information is not as critical as access to technology that allows one to get on the system to network with other people. We have to stop thinking of the public as an audience only. The public also provides information through small networks it creates.

Third, we have to stop thinking of the industry as just electronic publishers, but somehow, as electronic networkers who will electronically bring people together. Most of the people who call INFO LINE, which is strictly an information and referral service, do not know the information they need. It usually takes a long conversation with information and referral providers to find out what a person's real problems are, and what kinds of information is needed.

Finally, privacy issues are fundamental in any public information service. In the Santa Monica system, privacy is dealt with in a straightforward fashion. There is no privacy because no one is anonymous. Everyone is identified and conversations are considered public, as if they were in a city hall council meeting. Although users may have no privacy in the Santa Monica system, the situation is the reverse elsewhere. According to a federal court ruling in Austin, Texas in March 1993, "users of electronic bulletin boards have a right to confidential communications" (Smith, 1993). In his opinion, U.S. District Judge Sam Sparks ruled that electronic services publishers are protected just as newspaper publishers under the Privacy Protection Act of 1980 (see also the Electronic Communications Privacy Act of 1986). Thus, on every electronic bulletin board, private or public, privacy issues such as user identification and use of the system are important areas of concern for the future.

INFORMATION ACCESS STRATEGIES

One of the most important issues in providing citizen information services must be the development of strategies for bringing these services to the poor. Numerous studies have demonstrated that increases in the flow of information often widen the gap between the rich and poor (Donohue, Olien, & Tichenor, 1973). Information systems frequently tend to distribute products in a form that is most familiar to users who have more education, are more affluent and are therefore, the information rich. The information poor often see very little utility in much of the information that is available to them. Its very complexity is a barrier. Often they fail to see how the use of the information systems might improve their situation. Historically, although we have aimed to provide universal access, information has not adequately reached poorer rural or urban areas, contends communications professor Jannette Dates. One possibility involves providing free access to services in easily accessible public places like schools and libraries. These services could provide help in education, medical care, job ideas and skills training.

Another effective strategy involves putting information kiosks in places like grocery stores, shopping malls and taverns, where the public and private interests intersect. In fact, one Washington state grocery store owner was actually willing to pay for the system because he thought that if people came in to find out how to get a driver's license, they would be more likely to shop in his store than in someone else's.

Some of the solutions for improving access are quite simple and require no technological enhancements or improvements. At some information services, access can be severely limited by lengthy delays callers often experience. At INFO LINE in California, for example, callers must wait an average of 7 minutes before getting an answer. Many people hang up before their call is answered. INFO LINE cannot even keep track of how many people hang up and therefore fail to get food or shelter, simply because they could not get access to the information. Problems like these are not caused by constraints in technology. The solutions are often simply a matter of investing dollars in the needed places, such as providing additional phone lines or more staff support.

Computer literacy is another problem limiting access to information services. The solution, however, is a simple, nontechnical one. For example, in Harlem, computer literacy centers have begun inviting parents and students to come in after school and learn to use computers. These centers are providing the information poor with enabling skills to take advantage of emerging information services. Similar successful services are available in other urban areas, as well.

We can also easily deal with many language barriers by using multilingual kiosks and information and referral services. For example, one recent project of the Los Angeles Public Library successfully offered more than 200 social services over multimedia personal computer kiosks available in English, Spanish and Korean. We know how to provide multilingual services and have seen them work. There is simply not enough support to do it in the right way. From this experience, researchers such as Dutton, Guthrie, and Jacqueline O'Connell are beginning to discover some of the users that interest people from grass roots areas. These discoveries give them a better idea of what needs exist, and can help shape the ways that they design new information services.

LEGISLATIVE OPPORTUNITIES: THE AMERICANS WITH DISABILITIES ACT

In 1990, President Bush signed into law the Americans with Disabilities Act, one of the most widespread and sweeping civil rights laws in this country, according to Brotman, prohibiting discrimination against people with physical or mental disabilities. Although many citizens are familiar with its coverage of transportation, public accommodations, government services and employment, the Act also has a less known, but equally important connection to a national information service. Title Four, which is one of the major portions of the Act, deals solely with telecommunications, noted Brotman. Ultimately, it may have as significant an impact as any of the other sections.

Telephone technology has been available for nearly 200 years, and for almost 70 years we have had a universal service obligation. Yet about 10 % of the American population is cut off from what most people appreciate and perhaps, take for granted: the ability to find out everyday information. Twenty-six million people in this country have not had access to the services available on the telephone network because of speech and hearing disabilities. Thus, millions of Americans are unable to call the plumber, make a doctor's appointment, arrange a job interview, or easily find out about health-care options. That will change with Title Four of the Act.

Essentially the law mandates that a telecommunications relay service has to be provided for every American. In other words, if you are hearing-impaired you will be able to connect into the telephone network via a special device. A live operator will have a similar device and will then be able to translate that information to someone who can hear. For the first time, all Americans will be able to converse. We are going to be able to talk with people who have not been able to hear us and we are going to be able to listen to people who have not been able to speak. This is an important principle in terms of expanding universal service.

Several other provisions within the Act also have broad implications for developing a national information service. First, we must be sensitive to the needs of large and often disenfranchised segments of the population, such as those with disabilities or minority groups.

Second, we must take into account the dynamics of technology. To its credit, the Act recognizes that over time we will develop an advanced intelligent network to solve many of the problems of access and interconnectivity. Development of this network will be a gradual process. Leaving everything to the marketplace is unlikely to lead to universal service. Thus, until the network is developed, we have to create some interim solutions. Once that network is in place, we will be able to pull back from government regulation and let the marketplace play a more dominant role.

A third principle is the need to balance federal and state interests. In developing a national information service, the Federal Communication Commission and the individual states will play important roles. The Disabilities Act presents a very interesting opportunity because it charges the states with developing individual plans. In other words, there will be 50 different state laboratories for developing telecommunications relay services. Although the commission will oversee this development by providing minimum standards, each state and each region will have a say . This speaks very well in terms of notions of democracy and citizen involvement.

TELEVISION DECODER CIRCUITRY ACT

A second piece of legislation of equal importance, according to Brotman, is the Television Decoder Circuitry Act of 1991. Again, it points toward the idea of universal service. After 1993, this Act will mandate that micro circuitry be installed in every television set sold in the United States, regardless of where it is manufactured, to allow for closed-caption programming availability. This means that by the end of the decade, virtually every television set in the United States will have closed-caption capability. This raises notions of what role government should play in promoting technological innovation through industrial policy.

The Television Decoder Circuitry Act is beneficial because it mandates a very low-cost technology (i.e., the cost of a TV receiver will increase by only $10 to $15 according to Brotman) that will ensure that 100 % of Americans will have access to closed-caption programming. This affects not just people who are deaf, but also millions of people with bilingual needs or learning disabilities, as well as those who are illiterate by helping them learn how to read by seeing the words that they hear being spoken. Of course, these benefits will be limited to the programs that are encoded with closed-captioning.

INSURING PUBLIC ACCESS IN A MARKET ECONOMY

In designing information services, those who pay for the services often dictate who benefits. Thus, in our market economy, the private sector rarely creates information services that it believes will not be profitable. This could lead to problems of access and affordability. Similarly, services of great benefit to society may not be developed because they are not profitable.

When thinking about different models, we can define a set of services that are beneficial to society, and therefore, should be publicly funded. In terms of promoting access, public terminals are essential. For instance, the city-sponsored computer bulletin board system in Santa Monica gets about 20 % of its usage off of public terminals. This 20 % includes some of the most active users—even some homeless people. Research has shown that homeless people often use PEN terminals in public libraries to interact with others on the system.

In addition, we should focus on low-end technology. For example, 85 % of U.S. households can get touch-tone phone service and receive audiotext information services. Only 30 % of the people have computers; even fewer have modems. Although optical fiber may be needed to provide video on demand, such sophisticated technology is unnecessary for some of the information services being discussed here.

Finally, it is important to provide support for information services for the information poor such as Lifeline. Information services can offer an exceptional opportunity to connect city hall to the citizens it is supposed to serve. But carefully structured public policies are needed to ensure that services will enhance the lives of all citizens, not just the information elite.

The goal of universal service is premised on some notion of cross subsidy. Before the divestiture of AT&T, for example, the cost of local telephone service, which was considered an essential service (e.g., for emergency service) for every household, was subsidized by the revenues generated by long-distance calls. The idea is that by looking toward the greater good, we can determine how to put money into an overall pie and divide it back again. Of course, there have been real problems in how to create a subsidization program. The telephone company, for example, has been criticized for unfairly competitive practices for using cross subsidies to support the development of new services. A national information service may involve a variety of technologies such as fiber optics or cable television, making the notion of cross subsidies relevant and logical. Actual techniques for implementing cross subsidization may be different than what we have had in the past, however.

Another funding approach involves advertiser-supported systems. In this scenario, government applications can essentially piggy back onto the advertiser-supported systems with greater commercial viability.

Clearly, funding these public information systems and developing an adequate revenue base will be fundamental problems. But the variety of models now being tested, such as advertising-support or cross-subsidization, may provide an effective solution.

GOVERNMENT AGENCIES AS ELECTRONIC INFORMATION PROVIDERS

Local and state governments are very large repositories of information, which people either wish to obtain, such as library services, or they must obtain, like the tax laws and the rules for getting various licenses. Yet, all too often, extracting that information conjures up nightmares of being bounced around from agency to agency on the telephone, or waiting in line for hours to renew a license, only to get to the front and discover you were not told about some essential piece of documentation you had to show. Around the country, a number of creative local governments are using telecommunications to provide various information services to help citizens cut through red tape. Local governments have found that many of these services make operations more efficient and in some cases, can generate revenues.

These kinds of services are very good candidates for what Kendall Guthrie of the New York City Department of Telecommunications and Energy has called the "anchor tenets" of information services. Rather than seeking out an as yet undefined market, these information services prove they now meet a need because citizens are already using them to get information.

In the public sector, citizens and governments can improve their interactions in at least four ways. First, citizens need to receive some types of information from the government, such as license and tax information. Conversely, the government has similar information it wants to disseminate. For example, many citizens dial into audiotext systems when they call places such as the Department of Consumer Affairs in New York City. These systems have recorded messages answering frequently asked questions, such as rules and regulations for obtaining licenses, business hours, and the functions of various divisions. Although many people say they hate to get recorded messages, they probably hated getting a busy signal even more, which is what often happened. Now that New York City's Consumer Affairs Department put in an audiotext system, less than 30 % of callers get a busy signal compared to 50 % before the system was implemented. Moreover, citizens can get basic information about city services through these audiotext systems when it is most convenient for them and not simply during normal business hours. In addition to making it easier to get information from the government, these services ease the workload on government employees. The automated system can answer many commonly asked questions. This lets the city government workers focus on the more complicated questions.

Second, citizens want to retrieve a wide range of information from government beyond the information they need to have. For example, more than 50 public libraries around the country have their card catalogues on line, and allow people to dial into them. Users can dial in, see if the book they want is there. In many cases, they can even reserve the book, so it is waiting for them when they come to pick it up.

Third, cities are finding that their records are valuable resources. Many lawyers and title companies spend time in city hall looking through the real estate, tax, and land-use records. These companies are willing to pay to get that information from their own offices. Several cities have found that they can charge businesses a fee for dial-up access to this information and use the revenue to create a computerized database. This more accessible form of city records can then be made available free of charge to private citizens who come to city hall. Thus, the concept here is that you pay for the convenience of access, you do not pay for access itself.

Finally, we need to stop thinking about simply broadcasting information, and move into communication and transactions. For example, one of the exciting developments in New York City is the use of video conferencing to connect some police precincts with district attorneys' offices. If someone gets mugged in East New York City, reports the crime and wants to prosecute, that person eventually will have to talk to the district attorney, who is all the way downtown. If it is 10 o'clock at night, people do not want to get on the subway and go downtown. With video conferencing, they do not have to. Witnesses talk to the district attorneys downtown, via video conferencing, to collect the information. This translates into a higher prosecution rate.

As already mentioned, another useful technology for communication and transactions is the information kiosk. In California, there is talk about renewing driver's licenses or ordering a birth certificate at a touch-screen kiosk in shopping malls and other public places. The prototypes are already used by about 30 people a day. A number of these kiosks were installed in South Central Los Angeles after the riots. Up to 100 people used them each day to fulfill a variety of information and referral services.

FREEDOM OF EXPRESSION IN AN ELECTRONIC ENVIRONMENT

Freedom of expression in the U.S. essentially relies on two sources: the First Amendment and a diverse range of information providers. This combination protects the populace from government control of information and expression, and encourages citizen participation in the public dialogue. As we move into an increasingly electronic communications environment, legal questions over libel, copyright, First Amendment, privacy, reuse and other areas will arise. These issues most likely will be solved, as in the past, on a case-by-case basis.

This legal process will be messy. But even more troublesome, and potentially more damaging, is the lack of appreciation society seems to show for its right to free expression. Unless we recognize this right, we run the risk that there will be, at best, limited access and participation for a handful of privileged citizens. At worst, we will lose our freedom of expression altogether. We need to begin by identifying what First Amendment protections we now have, and what we wish to preserve in an electronic publishing environment. Then, we need to establish public policy that will accomplish that preservation.

In a country built on freedom of expression, public access and public participation, concluded Frank Blethen, publisher of the *Seattle Times*, "it is fascinating to see how the debate about electronic publishing is sometimes devoid of understanding of or appreciation for basic societal values. " Rather, the debate is often driven by technology and by corporate interests. For example, the debate often centers on who will provide information services. "Will it be the telephone company, newspaper companies or financial institutions." The discussion is usually framed in terms of the financial stake. Often debated is how many billions of dollars the new information infrastructure will cost. "What will it cost to lay fiber optics to the home?" How this new infrastructure will affect society and the democratic process is absent from the debate. This illustrates why we need to preserve these values in the future.

What are some of the important factors that should be part of this dialogue? Blethen argued that we must draw a distinction between electronic publishing and journalism. In providing news, he argues, journalism is based on a social service foundation. While also providing information, electronic publishing is more often primarily profit seeking, although one could clearly make this case about many forms of journalism.

Freedom of expression is protected by the First Amendment. This protection flows from the idea that providing news is a social service. Whether future electronic publishers enjoy First Amendment protection could hinge on whether the enterprise provides a social service, or whether it operates solely on a profit motive.

Much of the electronic publishing available today is merely an alternative means to receiving information that is already in the newspaper, according to Blethen. Instead, the reader receives the same information over the telephone, from a fax machine, or through a personal home computer. This kind of electronic publishing may or may not evolve to have some of the social service provided by newspapers or local broadcasters.

As electronic publishing unfolds, we need to recognize the public interest in news. We need public policies committed to preserving this vital social service. Blethen asks, "How do we distinguish between information that is socially useful, and therefore protected, versus information that is less useful to society?"

The people involved in developing news certainly want to make a profit, or they would not be in the business. But they also have other motivations, including the desire to provide a social service. Of course, electronic journalism is not new, points out Julius Barnathan of Capital Cities/ABC, Inc. The broadcast networks have been there since the beginning of television, and before that, in radio. Many broadcasters also own newspapers. Whatever the medium, some of the most profitable media organizations derive their success from strong local service.

Capital Cities/ABC, Inc., for example, is a multimedia company that owns newspapers and broadcast media. In its case, every station and every newspaper is locally oriented and commits extensive resources and energy to provide strong local coverage. Capital Cities/ABC properties, including the Fort Worth station and Kansas City newspaper, have won numerous awards for their efforts. The reason they do so well is because it is good business. Residents are happy with the product they receive and ABC makes money. ABC is number one in news, but also provides a public service.

Thus, it is not always easy to distinguish between journalism and electronic publishing. What makes them different? Is it the function? Perhaps electronic publishing encompasses nonjournalistic activities, like *Yellow Pages* ™ or selling cards? It appears that many of the people getting involved in electronic information are not necessarily operating from any kind of a social service notion. The emphasis on profitability is often stronger and clearer than that in news organizations. News outfits make money by selling advertisements, which are purchased based on the potential audience they will reach. In other words, news providers must sell their audience in order to attract advertising. On the other hand, many news organizations have entered the electronic publishing arena. As of March 1993, 465 U.S. newspapers had begun offering audiotext services such as news, weather and classified advertising (Glaberson, 1993). The *Washington Post* reports its 3-year-old Post Haste telephone system receives 800,000 a month.

In a national dialogue on electronic publishing, we often lose sight of the breadth and diversity of sources. It is easy to forget that most newspapers and broadcasters are local. Beyond serving a societal service function, news organizations are especially valuable to democracy because of this localness. In the newspaper business today, a current buzzword is *connectiveness*. It is a good buzzword because it recognizes that to be relevant, newspapers need to connect with their local readers. They also need to make their readers feel connected to the broader national and international communities. The point is that we need to make public policy to encourage localness and connectiveness as part of our social service foundation. The strength of the news media in our democracy is its breadth and diversity, qualities which must be strengthened today, and preserved for tomorrow.

Blethen contends that four very disconcerting trends are working against diversity in the media, and consequently, leading to less public inclusion. One trend is corporatization of the media. Short-term corporate thinking, Blethen argues, is widely recognized as one of the country's biggest economic problems today. Profit maximization is a byproduct of that thinking, and it ignores the public service aspect of news. Although newspaper publishers and broadcasters in America have always sought to make a profit, the 1980s brought profit maximization to an unprecedented level in the media as family ownership declined and public ownership grew.

The second trend is a growing concentration of ownership, adds Blethen. Too few people make too many decisions based on their short-term profit maximization and not on the local needs of their media. Too few news managers remain in local communities too short a time to foster any kind of connection, or to practice the social service aspects of news.

The third trend, posits Blethen, is the attempt by electronic service providers, government agencies and others to control content, access and transmission in the evolving electronic arena. The way to preserve our basic news social service values, and help ensure robust competition and participation, is to create a quality electronic conduit which is neutral as to content and access. To do otherwise is to ensure further alienation by the public, less public participation and a much higher risk that government interference will someday erode our First Amendment protections.

A final trend is that people are on information overload. They do not need more instant information, instant voting or instant polls. What they need is more mechanisms that allow people to meet, think, and discuss and debate information face to face. This leads to more thoughtful, and less emotional reactions.

All of these trends suggest that there is a distinction between news and information, and that certain information has a separate social justification. Should there be differences in the way we treat particular categories of information? In one sense, society should draw these distinctions in a way that protects the franchise of newspapers. The notion of free expression in an electronic environment presupposes the same tension between the government and individuals we see in the print and broadcast formats. Will this particular tension exist in the electronic environment, or will new relationships arise?

One of the really interesting questions is whether the press will persist in its current form. This has always been an important question under the First Amendment, which states that Congress shall not abridge freedom of speech and of the press. An important First Amendment question is whether there really is a separate entity called the press? If there is, what does it consist of? What is its social justification, and what are its responsibilities? The electronic environment may diminish this distinctive entity in some ways. In other words, we lose a notion of what constitutes the press. What does that mean for the First Amendment or for the society?

In 1990 there was the case of a computer hackers' newsletter called *Phrack* in which a youth got hold of a Bell South document dealing with a 911 emergency telephone network controlling police, fire, and ambulance services and published it over a computer bulletin board. It became a big freedom of the press case. The question was whether this newsletter, which was published like any other newsletter except that it existed only in computer networks, should receive the same protection as a broadcast or print organization.

This raises a variety of questions related to the First

Amendment. The first deals with copyright. What constitutes information that people can protect or have access to?

Another issue is whether electronic publishers have the same rights as newspapers when gathering news and protecting sources.

Fair use and copyright is a third issue. With greater resources and expertise in gathering information, will newspapers have an advantage over their competitors in the electronic environment? Do they have a right to use sporadically the information distributed through electronic networks? What will constitute fair use of digital images, text or audio, the original source of which may longer be known or identifiable?

Finally, what is the nature of libel in an electronic communication environment? If we are at a public meeting and someone says something slanderous, invades someone's privacy or says something obscene, the message quickly dies. On the other hand, if it is posted on a public bulletin board, where it is stored and distributed electronically, it is very much alive. Who will be held accountable in this environment? The creator of the message or the provider of the channel?

Santa Monica defines freedom of speech in the electronic environment strictly as a First Amendment issue. As owner of the system, the local government strictly limits its role when dealing with First Amendment issues. Comments are never censored. The local government does not eliminate personal attacks, threats and vulgarities, although they are not a major problem at the present. For the moment, local officials rely on self-policing. This is in contrast to privately run services such as Prodigy, the Sears-IBM consumer information service, which does censor offensive language and messages. How this issue is resolved is critical to the future of many of these public communication networks. Completely uncensored speech could turn these networks into a sort of speakers' corner, rather than a real town meeting. People may log off the systems and refuse to participate rather than put up with abusive communication.

Much depends on whether the transmitter has a right to control programming. Under principles of communications law, the carrier would not be held responsible for libelous communication, just as the telephone network is not responsible if a caller libeled or slandered someone in a telephone conversation, concludes Brotman. On the local level, it seems the sense of community will experience alterations and diminutions as well. In large part, the press reinforces the notion of community through the scope of its reporting.

But the electronic environment changes notions of boundaries, and raises the question of whether there are entities that have the capacity to control what defines the community. We often associate the First Amendment with notions of restraint and community. In other words, the First Amendment is about someone establishing regulations or controls about content. But it seems that one of the problems of the electronic environment is that it alters geography, both of the mind and certainly in terms of what constitutes a local community.

One of the consequences of electronic publishing may be the creation of what Cardozo Law School professor Monroe Price terms "diasporic utopias," a concept which has its roots in the 19th-century communities, had extraordinary cohesion and strict rules about behavior and participation. Rules focused on individuals' beliefs, what their work would be, and whom they chose as spouses. The West was spotted with these kinds of utopias, including the Amish and the Amana community in Iowas. These places were very cohesive, but were geographically bound. In a similar way, will the electronic environment allow secret or tight, closely knit communities to flourish in ways that are much more efficient than at the present time? What will be the limitations on rules for access to networks? Will managers be able to discriminate on the basis of race, or proven belief?

Barnathan identifies one of the other important consequences of an increasingly rapid and multichannel electronic communication environment. "People have immediate access to so much information they cannot digest it all. There is little time for analysis," he explains. This represents a growing problem for the audience, and an increasing challenge for journalists in the electronic environment.

NEW MODELS OF FREE EXPRESSION

Price articulates two means for measuring the achievement of democratic values in American society: the nature and quality of discourse, and the openess of forums to speak and to listen. Notions of openness, access and diversity are fundamental to effective democracy. The electronic highway can be implemented in ways that demean the nature and quality of public discussion. The illiusion of openness can mask the existence of greater concentration and tightness of access. A commitment to free expression seems to go beyond the mechanics of opening the market place.

Increasingly, we want to know what the information is. What is the substance of the marketplace? Is there, in some sense, the right kind of information? It is not enough to have free speech. We also must ask whether speech is being conducted in a public sphere. In sum, we might conclude that the electronic environment has the form to serve as a sphere of public discourse. But will it have the substance?

Acknowledgment: Research assistance on this chapter was provided by Josef Federman.

Part III

Policymaking Regarding Citizen Information Services

As we prepared for a national conference on the topic of citizen information services, we were all the more alerted to the importance of reviewing policy issues if public interest were to be pursued in our nation's future uses of these services, which together may constitute a new medium deserving of the attention given in the past national policy in broadcasting and telephony. Chapter 7 is a high level view of such policy implications from the then chair of the Federal Communications Commission, Alfred C. Sikes. The "national highways" analogy is complemented by chapter 8, in which Dordick and Lehman put the emphasis on the need for local connections to the national network, a need for "byways," so to speak. Chapter 9 summarizes the results of a panel of experts long experienced in communications policy issues as they address the idea of citizen information services. Finally, in chapter 10, we attempt a summary of what we have learned from this 2-year project, and to propose a course of action.

Chapter 7

CHARTING THE FUTURE OF COMMUNICATION SERVICES

BY ALFRED C. SIKES

This chapter, developed from Mr. Sikes' keynote speech to the national conference on citizen's information services, focuses on two dimensions of the so-called "information revolution." The first dimension is the idea of empowerment achieved through communications. In short, giving companies, organizations, and especially individuals more freedom, more flexibility, and more choice—more ability to understand, and shape the world around them. The second, and closely related dimensions, is the concept of democratization: that is, ensuring broader dissemination and ultimately universal access to new technologies—or more accurate, their capabilities. Mr. Sikes was chairman of the Federal Communications Commission at the time of presentation of these materials. He is currently vice president, New Media and Technology, the Hearst Corporation.

THE INFORMATION EXPLOSION

Some 50 years ago, Gertrude Stein observed that everybody gets so much information all day long, they lose their common sense. Can you imagine what she would think today? As I think about that, she was both right and wrong. I will first concede to her that today's blur of information in its dizzying and distracting tendencies has terminally affected some people's common sense. Take, for example, those people who are paying $50 to buy Madonna's latest contribution to literature, or more seriously, young people whose minds have been seized by television, or adults who let advertisers shape their self image. Unfortunately, short of unusual self-restraint or a constitutional upheaval, we will continue to be plagued by these problems. More encouragingly, however, we have the chance to use the explosion of information technologies—information and knowledge—to literally transform our society. Or perhaps more accurately, our society is already being transformed. Our challenge is to guide it in a constructive direction.

EMPOWERMENT AND DEMOCRATIZATION

We can better understand the challenges and opportunities if we focus on two dimensions of what journalists regularly label "the information revolution."

The first dimension is the idea of empowerment achieved through communications. In short, this means giving companies, organizations, but especially individuals—more freedom, more flexibility and more choice, as well as more ability to understand and shape the world around them. Second, and closely related, is the concept of the democratization, that is, ensuring broader dissemination and ultimately, universal access to new technologies, or more accurately, their capabilities. This seminal potential of communications and information advancement is not widely understood. This is especially true at the personal level. Information cannot take the place of knowledge, but it provides the building blocks of knowledge. So, putting more information in the hands of people will provide the building blocks that help our citizens become better informed and better involved people, who hopefully will make better educated decisions about their everyday lives and also about world events.

Perhaps the best way to explain the empowering potential of communications is to look back at some extraordinary numbers from the commercial sector. Between 1967 and 1988, the physical weight of U.S. exports, assuming constant value, fell by some 43 %. That is, $1,000 worth of exports only weighs about half of what $1,000 of 1967 exports weighed. This did not happen because steel got lighter. Information and skill combined to produce what often are called knowledge-based goods; these goods are increasingly the centerpiece of our nation's economy, and certainly of its export portfolio. That growing information intensity also may give you some insight into why in 1988, the United States only needed about two-thirds as many unskilled or semi-skilled workers as it did two decades before.

Perhaps even more telling, during roughly that same period of time, Japan increased the value of its industrial output by some 250 %, but without appreciably increasing raw material or energy consumption. While the Japanese trajectories have been the most dramatic, similar advancements have occurred around the globe. Whether you are talking about computer-aided design, manufacturing, or just-in-time manufacturing practices, or choreographing production on several continents, companies have been relying more and more on the intersection of computer and communications and perhaps more importantly, an increasingly skillful work force. That has been true both when it comes to making familiar products—cars, appliances, textiles—as well as when it comes to making new, much more advanced products such as G.E. and United Technologies' next generation of fuel-efficient, environmentally friendly jet turbine engines.

This new technological accomplishment is predicated on an entirely new and, a few years ago, unexpected level of knowledge. More intensity, more innovative use of information has freed business from many of the traditional limits, from the need to readily access transportation, raw materials or for that matter, an urban work force.

It has also contributed to individual freedoms and given more individuals more control over their lives. Now, if you're a South Dakota high school student, for example, you do not necessarily have to plan to move out of state to get an information-related job. Satellites and computers, and in the case of banking (Citibank) now offer you a home-based option.

Another obvious result has been to sharply increase the importance of using information systems and doing so well. Both today, and certainly tomorrow, the companies and people which are best at capitalizing on computers and communications will be the winners, that is, assuming they are using these tools to respond to the customer. In fact, these tools make customer-driven manufacturing increasingly necessary.

Before going further with the accelerating pace and intersection of knowledge, innovation and work, I want to examine democratization. That is, expanding personal opportunities. A little more than a decade ago the only telephone customers who got a large mix of equipment options (i.e., who were fortunate enough to be presented with a Chinese menu of service and pricing options) were what we would call national or flagship accounts. You know them, Sears & Roebuck, General Motors, or the federal government's General Services Administration. Little more than a decade ago, the only people who got to watch movies on demand were motion picture studio executives who had their own screening rooms, or maybe the President when he went to Camp David. The only people who had car phones were government officials and some phone company executives.

Twenty years ago, the only people who had almost immediate access to international news and events worked for U.S. intelligence agencies, or perhaps their counterparts overseas. But today, virtually all of these options which used to be available only to a small and privileged elite, are almost completely universal.

The new communication technologies are so commonplace that people do not even stop to think about them. Today you can get more kinds of phone equipment than I can explain. You can watch movies more than anybody would want to, and then you can watch them in your living room. And of course, we all watched Desert Storm unfold—patriot missile by patriot missile. Again, these are all virtually universal options.

Notwithstanding these extraordinary developments, I

know that we have just begun to see innovation. Twenty years from now, people will look at the 1990s with some bemusement as they ordinarily use the same capacity or bandwidth that major corporate users today take for granted. We will need breakthroughs at the personal access level. Even though Americans are becoming more and more computer literate, universal access will require generational improvements and user friendliness. We will also be applications visionaries. Educators, corporate leaders, health care providers and others, who can get beyond the ever-present barriers will truly reform the institutions. Business people have no choice; they have to compete, they have to adapt.

In government and not-for-profit institutions, leaders frequently see advocacy or politics as substitutes to adaptation. We are going to have to devise means of overcoming what seem to be almost an endless number of institutional, and often regulatory, obstacles.

POLICYMAKING IMPLICATIONS

Lawrence Peter of the Peter Principle, once quipped that government tends to defend the status quo long after the quo has ceased to be the status. Well, I certainly wouldn't lay all the blame at the government's door. Many, many established industries and institutions are caught in a Bermuda Triangle of inertia. In fact, we have been watching that unfold intensely in the shakeup at General Motors.

But rather than being pessimistic, or dwelling on the past, we need to look to the future, and the Federal Communications Commission needs to stay in the forefront. We, at the FCC, have taken up the challenge of encouraging and facilitating innovation, entrepreneurial endeavor, and new job and investment creations. Looking back for a minute, in the last three years we have established an emerging technologies bank of radio frequencies to make services like a new generation of pocket phones, and low power satellites technically possible. We've acted to eliminate regulatory barriers to competition. We want the telephone companies to push the cable companies and vice versa.

I am certain that the force of competitive pressure will speed the availability of a video dial tone. We want new satellite services to push each other, and the wire-based carriers as well. If they do, each will do a better job. Out of this competitive milieu will come creative partnerships—partnerships aimed at opening new markets and extending capabilities through the power of synergy. The Commission has also been working hard to ensure access for end-users, the consumers, information service providers, as well as competing networks.

We believe access, both at the wholesale and the retail levels, is quite important. Japan, a resource-poor nation, has led the world in developing information-based manufacturing. The United States, a customer-driven nation, has the chance to lead the world in developing an information-based consumer economy. As the commercial revenue stream brings us new products and services, the government and the not-for-profit sectors need to aggressively apply these new capabilities toward important social goals. In fact, as the combination of Chris Whittle and Benno Schmidt suggests, if the government-based educational sector fails to respond, we might be at the earliest stages of a commercial takeover of what today in education, has been largely a government or not-for-profit sector function. In my view, the transforming potential new knowledge technologies puts much of the government and not-for-profit sectors at risk. Adapt or perish may be the newest challenge, just as in the business community.

Many of the same people who say that Aunt Minnie only wants plain old telephone service, regularly challenge the vision of an information-based economy. I simply do not agree. With some 30 million personal computers and better than half the work force working with computers, we are rapidly becoming a computer-literate society. Certainly that is true of today's children and younger adults. In commercial terms, we're approaching critical mass, if we haven't already.

Electronic information services that use computer platforms are growing by nearly 20 percent a year. According to one industry estimate, the two most popular residential information services, Prodigy and Compuserve, have almost 2 million users between them.

These same factors which caused industry to adapt should cause American society to do so as well. For instance, one of the FCC's resident demographers always notes that if you graph the growth of the microwave oven, the automatic bank teller and mail order sales businesses against the movement of women into the general work force, you'll see almost a perfect match. All those enterprises, of course, are time savers. The one thing that all women who work outside the home have in common is that they place a premium value on their time. If you look at priorities in transportation and education, in energy, in land use management, what you will see is that far greater dependence on communications and computers will become the reality.

President Bush signed into law legislation, interestingly, sponsored by Al Gore, establishing a national research and education network. He noted at the time, the great potential that exists for many news services and new competitiveness gains inherent in this endeavor. He also praised the contribution which the U.S. computer industry has already made. But now, the concept of a high-capacity network needs to be broader. We need not only links between research and education centers, we need to link homes across America as well. If we follow through on this vision, in just a few years we should see a democratization of this potential and the empowerment of the American people who will have the potential to truly transform our society.

QUESTION: WHAT ABOUT CITIZEN LOSS OF CONTROL?

Audience question: Mr. Sikes, you mentioned the importance of increasing technological progress, democratization, and empowerment of people. What about the downside; that is, loss of control, constant monitoring of individuals, of their habits, the misuse of information to take advantage of others?

Alfred Sikes: My view is that, overall, the information technologies and especially the third and fourth generations as I see them, will give people a lot more control, not less control. Now, I don't think there's any doubt that we have a lot more data about ourselves in the general flow these days, and that it is more accessible, at least, on a theoretical basis, and maybe in many instances, on a practical basis. As we confront distasteful uses, we simply have to act either by rule or by legislation, to limit those uses. You brought up another point in regard to the intrusiveness of these new technologies. We now have legislation that essentially says that people have the right to put their names on "do not call" lists, just as we gained the right some years ago to put our name on a "do not mail" list. We are going to have to use those kinds of remedies when we confront those information technology uses that injure or harm important societal values.

In all, however, I consider the upside of technology uses far greater than the downside.

QUESTION: WHAT HAPPENS WITH A CHANGE OF PRESIDENTS?

Audience question: What happens if Clinton gets elected?

Alfred Sikes: Well, first of all, I won't be chairman. But I think, in many respects, there is a bipartisanship in communications policy making. Certainly, over the three and a half years, I've had a number of instances where I've worked closely with Democrats—and I think, effectively. There are also areas where there are probably some differences of opinion. I don't mean this partisanly, but I think it's fairly clear that there have been Democratic commissioners who have felt more interest in such things as the fairness doctrine, determining whether editorial judgment is being used fairly or not in the broadcast medium.

I believe no less in the fairness doctrine or the need to strive for balance, and I've spoken out on that. But I am a little reluctant to have the government begin to count seconds and look specifically at whether something is or isn't an adequate response to some fairness doctrine kind of obligation.

There probably would be a bit more tendency to want to be more regulatory. But I can tell you how that tendency plays out today at the FCC. The questions take longer. Because we have so much demand for the use of our organization and such limited resources, it's not going to get better. As we are told by Congress to do more and more, or as we of our own initiative do more and more, the questions just get longer and the frustration that results from that gets more intense.

QUESTION: WHERE ARE WE ON A HIGH DEFINITION TELEVISION POLICY?

Audience question: Next year, the FCC will make a decision about HDTV policy. What exactly will be the criteria the FCC will use, other than strictly technical? What is the vision of how HDTV will be used? Will it be more of the same that we're used to on current broadcasting cable (i.e., mostly entertainment)? Or, as some people have argued, should it be used more for information services, because that actually will impact the decision, the format that you choose?

Alfred Sikes: Well, my vision on this is really fairly simple. We have to optimize the developments in the marketplace and as a consequence, I think we now need to be looking at a digital system rather than an analog one. You might say, "gosh, that's old stuff," but, you know, a year and a half ago everybody was saying that there is no possibility of going digital. As the digital format became increasingly likely, I told the advisory committee to open up the multimedia work, create committees, look at extensibility, and scalability much more closely. And they've done that.

Obviously, picture and sound improvement will remain important. They are important in all media, but clearly they are important in the broadcast, VCR, cable, and DBS media; all will, of course, be users of the new technology. We must take a look at the interference properties of these systems.

There's also an economic dimension, although I'm not sure how big a slice that will be—namely, the implementation cost of a given system. The more complicated it is, the more processing-intensive it becomes both at the professional equipment end and at the reception equipment end. Both could have an enormous economic impact on conversion.

All of these things have to be and are being considered. It is a bit of a balancing act but overall, I have been very pleased. One thing I do not worry about is that we are coming up to a decision. I think that we have developed a great deal of momentum behind a very important technology. As you get closer and closer to making a decision, the centrifugal forces become greater and greater. I do worry a little bit that we're not going to stay with what we set out to accomplish, and I think that would be tragic.

Chapter 8

INFORMATION HIGHWAYS: "TRICKLE DOWN" INFRASTRUCTURE?

BY HERBERT S. DORDICK AND DALE E. LEHMAN

Increasingly we hear the analogy of a national information "highway" with the development of the interstate highway of some 40 years ago. But analogies only go so far. As Herbert Dordick and Dale Lehman argue, we must take care not to be led astray by concentrating at the national network level when the main challenge is to bring the advanced network and its services to the local level. If we wait for "trickle down" infrastructure, everyday citizens and institutions like local schools may be the last to benefit (if at all) from a national telecommunications infrastructure initiative. Herbert S. Dordick is a professor at Temple University, and Dale E. Lehman teaches at Fort Lewis College in Colorado.

ORIGIN OF THE "HIGHWAY" ANALOGY

The analogy between highways as infrastructure for the manufacturing age and telecommunications as infrastructure for the information age is compelling. Vice President Albert Gore, Jr. is, perhaps, now its most influential spokesperson; see, for example, his 1991 *Scientific American* article, "Information for the Global Village." The analogy is currently being used by the Clinton administration (Markoff, 1993). The comparison appears in most all telecommunications infrastructure studies, including the report by the National Telecommunications and Information Administration (NTIA, 1991): "Several parties repreated the oft-stated analogy that telecommunication facilities and services will be as important to the future performance of the U.S. economy as transportation systems have been in the past." Or as Kahin (1992) put it: "Is the Interstate Highway System precedent for the NREN? ... probably only that the transportation infrastructure as a whole compares to the emerging information infrastructure in terms of scope and heterogeneity."

It is clear that the highway infrastructure (in particular the interstate) was a great boon to the manufacturing based economy. It enabled rapid and inexpensive transport of goods and people throughout the U.S. It was supported by virtually all political constituencies. Users of the interstate include all segments of the population, although some are more intensive users than others. The sharing of costs is the major economic consideration underlying public support of the interstate. Perhaps the greatest economic criticism of the interstate is inefficient pricing structures (Small, Winston, & Evans, 1991). The two primary inefficiencies in highway pricing are (a) absence of congestion pricing and (b) inappropriate cost sharing among different user classes. For example, trucks are not priced according to the costs they impose on highway maintenance and construction costs that are related almost entirely to weight per axle.

In the narrow sense of a national highway infrastructure, the interstate must be deemed a public policy success. However, in the wider sense of transportation infrastructure, the interstate has been a costly failure. Car ownership is almost universal, but the importance of the interstate for different groups has been far from neutral. The role of the interstate in facilitating suburbanization is well documented. The neglect of mass transit alternatives and the subsequent impacts on residential and industrial location patterns is also well documented. The results may have been beneficial for wealthier suburbanites, but certainly have been detrimental for the urban poor. Urban America's problems are inextricably linked to the interstate. The interstate also facilitated the transition from small, heterogeneous, owner-managed shops to national homogeneous, corporate chain stores. Mass transit alternatives would have been more sympathetic to the former than the latter. The short term interests of suburbanites have been advanced by the interstate, but it is unclear that these are lasting benefits. Avoiding the costs of urban blight to everyone, suburbanites included, seems to be elusive over the long term.

The interstate also significantly altered the rural landscape. Small towns were by-passed as traffic moved from local roads to the interstate, eliminating many small businesses and often the towns as well. Strip businesses emerged crowding highway exits with fast-food restaurants, motels and gas stations but with few homes, schools and other necessities for community life. Rural America still suffers from the isolation created, to a great extent, by the interstate.

A complete cost-benefit analysis of the interstate is probably impossible, and is not our intent. Rather we wish to explore the underlying infrastructure theory which justifies the interstate as good infrastructure. We call this the "trickle-down" infrastructure theory. The most successful firms, industries and households were the primary beneficiaries of the interstate. At least part of their success must be accounted for by the availability and relatively easy access to the interstate. While poorer households, firms and industries did not benefit to the same extent, some of the benefits of the interstate trickled down to these groups as well. The postwar economic growth of the U.S. was, thus, shared by almost all groups. That the seeds for today's problems were also sown is only now becoming clear. The facilitation of inequities and the geographic isolation of certain socioeconomic groups now imposes a cost on us all.

A "WEAK LINK" THEORY OF INFRASTRUCTURE

We posit an alternative infrastructure theory—namely, a "weak link" theory. This theory dictates that infrastructure investment be focused on improving the efficiency of the weakest link in the economic structure rather than enabling the most able segments. Had transportation infrastructure been motivated by the weak link theory instead of the trickle down theory, there would have been more investments in mass transit and less in the interstate. Many of our most pressing societal problems would, arguably, have been reduced—environmental pollution, educational inadequacy, crime, rural economic stagnation and inequitable health care delivery.

The trickle down theory would appear to be motivating telecommunications infrastructure policy. The need for advanced services, new network capabilities and greater bandwidth can be traced to those segments of the information age economy that have been most successful. Indeed, the relatively recent use of the term "infrastructure" rather than "network" tends to support the need for a single, centrally managed and controlled electronic communications "highway" favored by the major carriers. It can be argued that by meeting the needs of the more informationally and technologically advanced segments ("the information literate") the benefits will trickle down to all through enhanced economic growth. The cornucopia of human services that will benefit mankind is overwhelming but first must come the electronic highway, the infrastructure, from which the manna will trickle down. Before we unquestionably succumb to this theory, let us realize that there is an alternative view of infrastructure. The weak link theory would lead us to address the needs of the least well off members of our society, those who are the least information literate, often poorer, less well educated, and least efficient. On what basis can this be sound public policy?

THE PRODUCTIVITY DILEMMA

Before addressing the foregoing question, there are a few disturbing facts about the information age to consider. Early observers of this age suggested that the driving force of an information society would be the production and consumption of information. Researchers in Japan and the United States, countries that count themselves among the leaders in the race toward becoming information societies, have attempted to measure the production and consumption of information. And while one may have serious reservations with the measures used in these two studies, the results are sufficiently comparable to teach us something about the road to informatization. In both countries the quantity of information produced, via mass media and telecommunications, grew almost three times faster than the quantity of information consumed (de Sola Pool, Inose, Takasaki, & Hurwitz, 1984). Information can be produced at a very rapid rate, but it can be consumed at a relatively slow rate. We have not developed the means nor the capacity to efficiently select and organize information for our consumption, hence, "information overload" and "information anxiety" (Wurman, 1989).

Robert Solow (Schor, 1992) has pointed out that computers can be seen everywhere except in the economic statistics. Productivity has not improved, especially in the service sector where information technology spending has been most intense. Productivity in this sector may very well be the cause of the stagnation of overall productivity in the United States which has been a major cause of the exploding national debt and the nation's inability to compete in many world markets. Yet, people are working longer hours than ever before; unemployment among information workers is high and will probably remain so for some years to come, and the workplace has not been humanized. Automation and computerization on the factory floor was supposed to release workers from the tyranny of machines and Taylorism. In the past 20 years since, the years during which an information economy fully blossomed in the United States, working time has steadily increased to the point where the average worker now puts in an estimated 168 extra hours of paid labor a year, the equivalent of an additional month of work per year. Almost 40% of households in the United States have dual wage earners, usually an employed mother who works an average of 65 hours per week, including housework, child care, and employment. More than 30% of fathers with children under 14 work 50 or more hours per week. Computers seem to control workers in a subtle way, monitoring workers quietly and persistently.

Hard economic times no longer affect only the blue collar, the unskilled or semiskilled workers. Middle managers have been hit the hardest as computers and communications have usurped their roles. Corporate downsizing rather than employee re-training to increase productivity has become the key to profitability in the 1990s and relatively high paid jobs have given way to much lower paid information workers. Information work is not always high valued work.

Even with computers in the workplace, it takes 4-6 weeks to change an address for a magazine subscription. Making a simple purchase at a computerized store is frustrating, taking too long and often resulting in errors. Technology has given the uneducated and unskilled the opportunity to have low paying jobs with no need to enhance their skills. Part of the cost of this inefficiency falls on the customer. And part of the cost is external to the transacting parties but appears through the evils of perpetual underdevelopment. A rising standard of living requires high valued work performed by a skilled workforce. By all the usual measures, the United States is an information society, yet we do not have a workforce capable of taking us into the next century. The Los Angeles riots were a warning about what happens when infrastructure fails to address the needs of the masses.

Let us bring these arguments to our doorstep and into our homes, as an example. One of the authors recently went through the trauma of purchasing a new home. An incredibly long and broad chain of people was affected by the mortgage application. Upstream were the sellers of the author's new property (dependent on the author's closing for their down payment) and the original seller of this house (who had provided a swing loan and now required the cash to pay for the contractors on the buyer's new house.) Downstream were the several contractors the author had aligned. Then there were the purchasers of the buyer's old home. Home buying what it is, everybody's productivity suffers when the process is delayed or its outcome uncertain.

In our increasing specialized and interdependent economy, a vast number of people's productivity was impacted by the author's mortgage application. The application, in turn, depended on a large number of other people, namely, bank personnel, mortgage company personnel, creditors, employers, and companies who compile documentation. The technology available in the U.S. to this chain of actors is unsurpassed. The trickle down theory would suggest that such mortgage application processes are more efficient than ever. The author's experience suggests otherwise. Incompetence was evident at all levels. The form it took was simple—misfiled papers, misunderstood documents, incorrect and incomplete addresses on important mailings. The weak link theory says that the efficiency of this entire chain depends on its weakest link. That link is often an underpaid, unmotivated, unskilled worker with neither the incentive, the ability nor the training to work efficiently. They may even lack the tools to enable them to work more efficiently, especially if their firm is a small one. Of course, had the author been wealthy, these problems could have been circumvented. But the number of people in such an advantaged position are few. Most people must still apply for mortgages and engage in other chain transactions. The ability of these clients of the transactions, as well as the productivity of this vast chain of affected people, then depends on this weakest link.

The U.S. workforce is dividing into an educated elite, with brainy and computer-connected lawyers, investment bankers, consultants and scientists and technicians on one side; and vastly more people without college degrees toiling at the only other jobs—ringing up groceries, cleaning office buildings, part-time fast food jockeys and making deliveries. Increasing computerization and automation of factory work is eliminating the need for skilled and semi-skilled factory workers. The trickle down infrastructure increases this gap by further enabling the former to be productive and offers little to the latter. The weak link infrastructure would enable the latter to become more productive and will, thereby, increase the productivity of all. Improving the efficiency of this weakest link would "trickle up" to this chain of more skilled workers.

A WEAK LINK TELECOMMUNICATIONS INFRASTRUCTURE

The importance of the weak link theory is that it produces markedly different infrastructure policies. The telecommunications infrastructure cannot be divorced from the larger information infrastructure. What does the weakest link in the information age need? The goal becomes ease of access and use for consumers and ease of entry for information providers in the form of their personal computers. Infrastructure aimed at these criteria cannot avoid dealing with the issues of education and training (or retraining), terminals and user interface standards. For example, the computer literate already have terminals. The masses without any screen based device account for over 73% of U.S. households. Of the somewhat fewer than 30% of households with personal computers, only about 5% are equipped with modems and fewer are actually on-line (Dordick & LaRose, 1992). Just as the interstate benefited those who already owned automobiles, the early beneficiaries of NREN will disproportionally benefit those already owning computers—unless an alternative policy course is adopted.

Most of the state telecommunications infrastructure studies (e.g., Tennessee, New Jersey, and Pennsylvania) have boiled down to recommending the increased speed of fiber deployment. Attention solely to bandwidth is misplaced. It is not that fiber-to-the-curb (or enhanced compression technology as a substitute) is not important. The potential for benefits are clear: Medical experts and high quality education, among many other services, can be delivered to even the most remote locations (e.g., Fisher, 1992; Williams, 1990). But, bandwidth alone will not enhance the viability of small business at that remote location and bandwidth will not create the incentive or ability to upgrade skills. On the other hand, universal provision of terminal devices could facilitate such development. We are not suggesting the French Minitel approach for the United States, but that focus on terminal devices may be more appropriate infrastructure policy than focus on network bandwidth. Nicholas P. Negroponte's (1991) analysis of the substitution of intelligence for bandwidth is pertinent here. Most people currently have terminals with the minimum of intelligence, voice telephones and television receivers. If terminal intelligence (PCs) were universal, much bandwidth could be substituted and the need for the remainder would be market driven. "Dumb" terminals are an intermediary step. Focus on terminals would ensure that the needs and abilities of the weakest links be considered. Focus on bandwidth circumvents the weakest links in a way that could significantly weaken our interdependent information age economy.

The highway telecommunications infrastructure analogy provides insight into the benefits and costs of alternative public policies. There were at least two approaches to providing a transportation infrastructure as there are now at least two approaches to providing an information infrastructure. If we choose to provide telecommunications "highways," then we will further segment society. If we keep our focus on information highways in the broadest sense of ease of access and use for consumers and ease of entry for information providers, then telecommunications could be a key for economic development.

The tension between the trickle down and weak link theories is evident in the evolution of the NREN. It originated as NRN—the National Research Network. Congress added the "E" for education. The NREN as the National Research and *Education* Network, is mandated to develop a platform for the advanced needs of the research community, as well as enhancing access to existing and new services, even for K through 12 schools. A quick look at the appropriations under the legislation shows definite bias towards advanced research needs rather than broad access. Yet access by schools and small business are clearly in the national interest at a time when education and re-training are high priorities and for an economy in which small business accounts for major growth in jobs. The political attraction of high visibility state-of-the-art projects is clear. The marginal benefits of such projects vis-a-vis public access promotion is less clear.

Highways facilitated the automobile and oil industries among other mostly large businesses. In the retailing area, large chain stores could capitalize on the economies of scale in advertising which highways permitted in a way that small stores could not. Pedestrian traffic was much more suited to small scale retailing. Telecommunications highways may similarly benefit large "infotainment" enterprises (e.g., CATV or media conglomerates). Bandwidth can easily meet the needs of such organizations with the internal capabilities to use it for their advantage. Terminals more easily meet the needs of the average small business. From an economy-wide perspective, the greatest marginal productivity gains may well lie with small businesses. Large conglomerates, often less susceptible to enhanced efficiency through information technology, are often not well served by the public network and are increasingly foreign owned. They are also prone to entering into arrangements for private leased lines or even to installing their own networks.

Consequently, local exchange carriers are witnessing the loss of their high volume customers but remain responsible for an obligation to serve those that remain. They are under pressure to upgrade their public-switched networks to meet the needs of the more advanced users while still fulfilling the social mandate to serve all at reasonable (low) rates. These pressures may prove insurmountable as "private" networks pick off the more lucrative subscribers, leaving public networks unable to afford to upgrade. The result will be an even greater distance between highly paid information literate workers and low-paid information illiterate workers. The public policy questions are: Does such uneven development serve the public interest? Will the benefits trickle down, enhancing wealth overall? Or will the weak links hold us back from realizing the productivity gains promised by the information age? Could the information rich actually pay less under a weak link approach than a trickle down approach? They would forego the immediate benefits of their own enhanced capabilities, bear some of the costs of weak link policies but would gain the benefits of upgrading the weak links. Ultimately, productivity could improve by permitting specialization that works and the subsequent costs of supporting the information poor would be reduced.

The ultimate economic question is: If the weak link theory approach to infrastructure is more efficient, will it be achieved by reliance on market forces? There are several reasons for skepticism. The nature of weak link benefits is that spill-overs are endemic. It will be difficult to contain the benefits so as to recover the considerable costs of market creation. This is important since weak link approaches are likely to be expensive. The cost of a terminal policy that includes education and training will be high. In a fragmented competitive environment it will be difficult for any single firm to recover these costs. There will be many network externalities and agglomeration economies that may be difficult to capture. Any attempts to earn revenue from terminals and use of services will be prone to "bypass" in the competitive environment that now exists. For example, if the local telephone company were to invest in weak link policies and recover the costs through surcharges on information service access, they could easily be bypassed through the use of other local telephone numbers (which carry zero marginal usage prices in most of the country). One could easily dial an information provider with a free local call on a telephone company subsidized terminal, thus bypassing the telco surcharge. There are mechanisms which could prevent this—for example, a local measured service for all calls—but the general problem remains. The investment costs are substantial and "taxing" the benefits will be difficult and possibly inadequate to finance these costs.

There are risks to the proactive policy of providing terminals, training and education as infrastructure. Speeding up the market development may cause problems for private parties whose financial well-being depend on a niche approach. This "tyranny of the niches" causes most consumers to lose. It may, indeed, be a waste of taxpayer money and it may cause adoption of obsolete technologies. But a case can be made that with many electronic highways (cable and the telephone—who said there must be only one highway?) a government proactive weak link approach to infrastructure might bring order to a market sorely in need of order. It would force compatibility of the multiple highways which may very well result in dramatic cost savings. If digital cable television becomes a major player in the information infrastructure, and, it is very likely that it will, consideration must be given to common carriage principles regarding standards. To date, regulations have been concerned solely with price, thereby repeating past mistakes of permitting the extension of market power to manufacturing and to new information services offered through "smart" converter boxes. This vertical integration limits access to sources of information and may also limit the development of multiple providers of information. Furthermore, ease of access to universally available and compatible networks facilitated by proactive terminal and training policies will create a mass market for information.

FUNDING A WEAK LINK INFRASTRUCTURE POLICY

Income tax credits for terminal acquisition might be an appropriate way to fund this alternative infrastructure policy. We can learn a great deal from the French Minitel experience and avoid their mistakes. For example, 15-20% of Minitel subscribers never use their terminals and terminal technological obsolescence has been a problem. But too much has been made of this. Major shortcomings such as the slow display speed have often been overcome by software adaptation permitting changing small parts of the screen rather than requiring an entire screen to be redrawn. Tax credits would probably rule out most of the non-users since effort is required to obtain the credit. An advantage of tax credits is that an industry/regulatory forum could define a "basic communicating terminal" in order to facilitate technological flexibility. Defining this "basic terminal" will also facilitate the multi-mode issue, cable television-based devices, personal computers, dumb terminals and wireless devices could also be included in this industry/regulatory forum. Technological obsolescence could be minimized because this basic terminal could allow for modifications and upgrades to meet the increasing skills and the requirements of the information providers as they compete for customers in the information market.

If such a policy were implemented, it is likely that fiber would be quickly deployed in both telco and cable television networks. The logjam in the local loop, fomented by debates on how to fund network upgrading, could very well be broken. Telcos would not require subsidy since the market for information would open rapidly because of the essentially universal availability of terminals and the potential for a much larger consumer market than would be available if only network bandwidth were expanded. The debate over who should pay for network upgrading would be moot; it would simply be good business to upgrade.

Assuming that a basic communicating device or terminal can be sold for between $75 and $100, universal penetration for the 100 million households in the United States can be achieved for between $7.5 and $10 billion dollars, a remarkably low cost for such a profound experiment. It would make sense for the Departments of Labor and Education to share this cost since they will clearly benefit, given the goals established by the Clinton-Gore Administration.

There are, certainly, other approaches to the practical issues of funding the weak link approach and they need to be addressed. We have managed to deliver on the promise of universal telephone service at affordable costs. Why cannot a mechanism be devised that will lead to universal information access at affordable costs? Perhaps this could be done as part of an enhanced telecommunications subsidy program that builds on Link-up America and the Universal Service Funds.

The market mechanisms may fail to reveal the relative efficiency of the weak link theory. What is troubling about the current infrastructure debate is that the weak link approach to infrastructure is not even being considered as an alternative to the trickle down approach. The latter is well suited to market forces where specific high valued information services are offered to a niche market with sufficient willingness to pay to make it profitable. If network and agglomeration economies are sufficient to warrant the more substantial costs of a weak link approach, there is no assurance that the market will reveal this. Indeed, the only experiment in that direction is Teletel in France which has not been adequately studied but has been overly dismissed as public policy. One of the important but often overlooked benefits of the Teletel system is that it made a great many French citizens computer literate. Regulators are all too quick to leave it to the market to develop information services and some states have codified this in legislation. For example, Colorado would treat information gateways as new services, by definition unregulated. There is little the Public Utility Commission could do to follow a weak link approach unless competitive markets were to provide it. It seems that by default the only infrastructure alternative is the trickle down approach. This makes the bifurcation of our society more likely. It offers the prospects of a mass market of information poor without skills or resources to deliver the promise of the information age. It proffers a Faustian bargain for the next generation of the information rich—to use their highly productive talents to support an aging population burdened with a mass of underproductive workers with whom they must share a society.

There may be no suburbs to escape to.

Chapter 9

POLICY IMPLICATIONS FOR CITIZEN INFORMATION SERVICES

BY ELI M. NOAM, CHARLES D. FERRIS, AND EVERETT C. PARKER

Some would argue that if we were to have as major a policy initiative as a nationwide citizen information service, this would require a Federal policy that would encourage standards, interconnection of regional, state and local information services, definition of a balance between public service and commercial services, and a means for financing the initial development of the service. In this chapter, three policy experts discuss their views of national policymaking. Their remarks were edited from presentations delivered at the October 27 national conference, Media, Democracy and The Information Highway. Eli M. Noam, a professor of finance and economics, directs the Columbia Institute for Tele-Information at Columbia University; he has also served a 3-year term of the New York Public Service Commission. Charles D. Ferris, an attorney with the firm of Mintz, Levin, Cohn, Ferris, Glovsky and Popeo, served as chairman of the Federal Communications Commission during the Carter administration. The Rev. Everett C. Parker is in residence as a senior researcher at The Donald McGannon Research Center, Fordham University.

ELI M. NOAM: THE POLICY CONTEXT

What can you say about policy and a national information service? In one respect, it is a term of enchanting vagueness and trendiness. It is also a Rorschach test for a world full of people who say we should do something in the communications infrastructure. But now there is a smell of real opportunity in the air and this brings the best out of people as well as the worst. Yet, we are astonishingly unprepared, which is why this gathering is very useful.

The beltway community—never of a particularly contemplative nature—is occupied with playing musical chairs, as the major law firms exchange their offense and defense teams to play government for the next 4 years. Much hype surrounds new services and obscures rather than enlightens. One has to be realistic; can any new administration truly overcome the accumulated unsolved problems of the past before they add their own?

We are dealing with a structural problem: We have gone from the old telecommunications regime, based on monopoly, to a new regime, based on competition. This inevitably creates all sorts of new problems. In the old days, AT&T basically took care of national planning for telecommunications, got an assured rate of return, and determined the technology. Today we do not have that environment anymore. Competitors and other industries, especially cable television, have established a rival wire line network system. Yet in some ways we have the worst of both worlds. We have competition but it has become more one of a political competition before government bodies rather than competition in the marketplace.

Now, we also have NREN (National Research and Education Network), whatever that is. Lots of people have been persuaded by the interstate highway analogy. But NREN is not a construction program. It is not about investing in fiber; it is a logical software-defined network. It's not an Apollo program.

NREN is aimed at the leading edge user such as university computer centers. As far as regular users, such as schools, and the like, are concerned, it's strictly a trickle-down technology. To address this, then, Senator Gore sponsored in the last Congress a legislative initiative that deals with broader applications and uses.

But for all of its vagueness, NREN is a good model. Let's look at its close relative, Internet, which integrates numerous sub networks, private sector firms like MCI and IBM, and nonprofit institutions like universities and lots of government agencies. Internet holds them together with some government money; not much, but enough to serve as a catalyst. Now, who is not involved? Lo and behold, the traditional government regulatory agencies, like the FCC and the state commissions. They hardly know what this is all about. I served for three years on the Public Service Commission in Albany and the word "Internet" was not uttered once.

Nor have the traditional carriers been much involved, and certainly not the cable television industry. All of these have been enmeshed in the traditional battles, whereas Internet and NREN have been somewhat of an end-run around the traditional system by the computer industry—which seems to be much better at working things out, perhaps because their ratio of technologists to lawyers is high.

Given these circumstances, what is likely to happen? First, there will probably be less ideology than in recent years.

Second, there will be more use of the government's procurement function. Government is running a huge communications network called FTS 2000. Yet, FTS-2000 was never contemplated in terms of economic and technology development. Therefore, I want to propose an "FTS 3000," which would involve government as a user of leading edge, and applications will involve national laboratory and the defense industry. It will also help government decentralize its functions closer to the grass roots. Samuel Morse's telegraph was supported by federal government money. In recent years, government was behind the important technology of packet switching.

A national communications grid is not so much a construction program but an interconnection arrangement to overcome the barriers between the separate network systems. It is clear that one needs to bring cable television networks into the communications infrastructure, and similarly also involve the other alternative local exchange companies. This is critical because right now, whenever Washington talks about upgrading the infrastructure, it is in terms of doing something for the telephone industry. Automatically, competitor industries oppose it. Therefore, a policy of inclusion, of establishing cable television as well as other rival networks, as part of a national network of networks is essential.

To do this, we need a historic "grand bargain" that would get some of the endless and mindless regulatory problems off the table. The various industries all want something. We have seen this in the context of the recent Cable Act, and it is possible now to fashion arrangements that could well move the agenda.

This will include, in the long term, a reform of the system of subsidies. The traditional system has been based essentially on a monopoly supplier where some customers would support other customers. In a network of networks environment one cannot do that anymore. A different system might be based on a universal service fund as the source for subsidizing certain functions and users. The money might come from a general communications value-added tax as a substitute for the existing hidden tax system.

Lastly, improve the administrative process. Right now, much of the process works in the context of an adversary model which takes you only so far. Chairman Sikes did a good job in moving the agenda along, but there is the problem of the FCC being used for dilatory purposes.

To conclude, the issues in the 1980s were those of opening and liberalization. It's important to recognize how successful that policy has been and Charles D. Ferris, in fact, was one who started it. But the 1990s will be marked by a different form of orientation, still continuing the policy of opening, but also assuring forces of integration to deal with the centrifugal forces that we have unleashed. Those directions will be characterized by "inter" words like interconnection, international, integration, and intermedia—to provide some glue to hold the system together.

CHARLES D. FERRIS: TOWARD A POLICY OF DIVERSITY

If I remember correctly, the question of this discussion is: Should there be a national telecommunications policy? Keep in mind that when the Communications Act of 1934 was enacted, there was a very strong bias in that Act to insulate the executive branch of government from communication policy. Congress felt that communication policy was too sensitive to the country and to the individual citizens to have potential abuses by the executive branch.

That is why Congress set up an independent regulatory agency, the FCC, which is really a creature of the legislative branch. They are not part of the executive branch: it is beholden to the Congress and those biases still exist to a great extent today. Bear in mind, too, that if we turn to government for policy about a national information service, there are questions some raise about rights of privacy, the problems inherent in the assembly of information, and the access to that information. Some people feel, I think rightly so, that probably the greatest potential abuser is government itself. These are biases that are a part of our culture, but also I think, are probably a very healthy part of our culture.

That being said, the very fact that 55 % of the American workforce is in information intensive industries—and the percent is growing—is a reason for information policymaking. Because the productivity of our economy will be determined in great part by the efficiency that information is collected, moved and processed is a reason that the government should be involved. This is the basic structure of our economy now.

But the bias is still there. I can remember my last year as chairman of the FCC, back in 1980, I was meeting in Europe with representatives of 14 countries. We were planning facilities across the North Atlantic, which turned out to be the first major undersea fiber optic pathway. We had many, many meetings to try to come to a common understanding on the project, then eventually we had to decide when to meet again to put the final bow on what was going to be a decision to approve the project. The chairman of the European delegation, a Swedish telecommunications expert, kept insisting that the meeting be the last week in October. This was October of 1980, and it was seen to be inconvenient to most people. I had expressed total indifference as to when we would meet. As it turned out, I found out afterwards he wanted that date because it would come prior to the elections in the United States. He felt that this was going to be a very positive telecommunications announcement and that it would be perceived as something very, very favorable to me and to the administration.

It really made no difference to me since telecommunications policy was unimportant at that time in the United States. No one cared about telecommunications policy. The *New York Times* would probably not even care that we made this particular announcement because telecommunication policies were so invisible at that time. I think that this was really because the telecommunications infrastructure that we had in the United States was so good.

We assumed good telecommunications here; in Europe they did not. They had terrible telecommunications; it was controversial; it was a political issue. Whether you would have good facilities or not was something that mattered politically in Europe. I was politically so naive that when the chairman of the cable delegation came up to me afterward and said, "I was trying to help," I said, "I did not even realize it, you know, thank you very much, but I don't think it would be a blip on anyone's screen in the United States."

That was back in 1980. Things have happened since then, and many things happen for the wrong reasons. What happened is that the United States Congress and the United States government have gotten interested in telecommunications policy over this past 12 years. To a great extent, they got interested in it for reasons of perception. The Congress has perceived that the FCC for the past 12 years has been ideologically driven. FCC Chairman Mark Fowler used to admit that he felt there was no difference between a toaster and a television set, and government should be interested in each about equally as much.

This upset some members of Congress because they felt that telecommunications policy, and certainly broadcast policy, was something that was more critical than a toaster. And so, they started getting interested in telecommunications policy because they did not trust their creature, the FCC, for ideological reasons. Congress has spent the past 12 years involving themselves in the details of telecommunications policy, and of course, the fruition of that interest was the Cable Act that was passed over the President's veto just a month ago. That was the first piece of telecommunications policy that has been enacted by the Congress since 1934 over the objections of an industry. Now Congress has an interest in telecommunications and a lot of people have been educated about it. They have found out that this is pretty exciting stuff to deal with.

How will this new interest by Congress manifest itself? I think we are at a very critical stage in telecommunications policy; the FCC over the next four years is going to have some very significant opportunities to determine really what is going to happen. To a great extent all of this is technology-driven. When I was at the Commission, they thought that we were doing all sorts of marvelous things in freeing industries from the shackles of regulation. But it really was the technology that was driving it. We were sitting on top of this tremendous kettle, and all we did was let it vent itself to some degree. You can try to contain all the steam in the kettle for just so long, but we just sort of let it ease out, and let the technology free itself. This removed some of the burdens that had been imposed primarily by lawyers in the prior 30 years, mainly to protect particular vested industries.

As a guiding principle of telecommunications policy over the coming decade, I don't think that government should pick winners, but maybe it should permit losers.

If anything, there should be a bias toward diversity in both conduit and content.

There is much dialogue going on now, as people interest themselves in telecommunications policy, about a "pathway to the home." I hate a pathway to the home because that means there is going to be only one provider of the conduit. We have a structure now that doesn't have a bias that way, yet people say it would be wasteful to have duplication of facilities. Different experts are talking about $500 billion over the next 20 years to build the telecommunications highways of the future—all of which seems like an awful lot of money. But if you recognize the fact that telephone companies and cable companies spend $20 billion each year on infrastructure improvement, then over 20 years, there is not much new money that has to be added to rebuilding the national telecommunications infrastructure.

If there continues to be the diversity of conduit that we have now, then you are going to see much more. Coming on stream now is about 300 megahertz of spectrum that is going to be made available for radio telecommunications at the local level. There is about 220 set aside for the new technologies; there's already about 70 megahertz for cellular telephone. There is about another 30 for paging and SMR, and there's talk about 200 megahertz from the government side that ultimately can be freed up. You are going to have more radio spectrum being able to make that last drop into the home, or directly to the consumer. We talk about one wire, but it is not going to be one wire. It's going to be either one wire or one radio link. It can be wireless cable. It can be cable by wire. It could be cellular; it could be PCS (Personal Communication Services). There are all sorts of options and when that happens, the last mile, or the last drop, will truly be competitive. By the fact that you have competitive pathways, then we are going to have a very healthy structure in the United States.

There is going to be multiple access for all the information providers, and it is very fascinating. There are about 12,000 information providers. I thought it was nine, but I was corrected. Still, I don't know what they provide. I mean, I know what LEXIS or NEXIS provides, but there are people who provide surfing news. Entrepreneurs are providing all sorts of things and it is a multibillion-dollar industry still in its infancy. The thing is, if there is a market, maybe these folks were running a Baskin-Robbins five years ago, but if they think they've got an idea, we'll see them in the information business. Good luck to them. Some might fail, but there will be others that come along and do it, just as long as they are not prevented from having access to consumers through multiple gateways. I think that is the most critical thing. If you have only one provider, and that provider provides both content and the conduit, we will be in deep trouble. We will retard significantly the development of a telecommunications infrastructure in the United States.

Let me say one more thing about the notion of a competitive environment. You have all heard that when they broke up AT&T in all the confusion, people hated it. I liked it when I got just one bill. Now I get this mess of bills—I can have MCI or Sprint or anything. It was so nice when I got only one. Even with the breakup, we have the best telecommunications infrastructure in the world. The best proof of that, in my mind, was when AT&T was divested by their own consent, back in 1984. One year later, Japan—a country that is very comfortable with hierarchical forms of organizations, very comfortable with monopolies—what did they do? They mirror imaged, by legislation, what we did here in breaking up AT&T. They legislated competition to NTT in Japan. The reason is that they knew the system that we have created here was going to be so much more responsive to emerging markets. They knew that with the U.S. telecommunications infrastructure open to competition, we were going to move far and away ahead of anyone in the world.

The Japanese were not going to let that happen. They were going to try it themselves although it went against their culture, really, to a great extent, because they have tremendous comfort with monopolies. But they did it. I think that is the best argument that what we have here is by far the best telecommunications system, and as long as we maintain a diversity of conduit and a diversity of content, I think no one will catch us.

EVERETT C. PARKER: REINFORCING THE CONSUMER VIEW

I am a member of the only group—I don't say I represent it—but it is the only group that really has an interest in free, open and robust competition: consumers. They may get upset at their bills, but they don't know how to protest. Consumers certainly need a communications policy and they need a certain amount of protection.

Now, the policy issues are not just speed of transmission and the ease of access to sophisticated equipment that those of us in this room, most of us, have. However, I would certainly support Senator Gore's effort to hitch up all of those supercomputers.

But our policy should be basically a three-fold thing. The first is to promote the underlying premise of the First Amendment, which is diversity of sources of information and guaranteed access to all sources. The second element in it should be privacy. And the third should be universal service.

If we are going to have this broadband communication highway, then every household and every small business should have open access to it. I am talking about universal service in a little different way than Eli did, because, in theory we have universal service now. It's in the Communications Act, but it's something of a myth.

In Watts, on the west side of Chicago, in Crown Heights, in Washington Heights (the latter, which you can walk to from Columbia University), there are no faxes, no voice mail or data processing, or even telephone service. If you try to use a pay phone, someone is liable to shoot at you because all the pay phones are taken over for drug dealing.

So, we have a very difficult problem. The way in which we are deploying and using electronic technologies makes it absolutely necessary that we face up to the moral and political aspects that a large segment of Americans are going to be kept from communications services that are available to all the rest of us. I think, to a very great extent, that is the most important policy issue that we face. Not the economic one. I certainly agree with Mr. Ferris, that the economic one can be phased out over the years.

Now, what do people really want? I did a study that encompassed the leadership of over 400 public organizations and religious organizations, trying to find out.

We tried to explain to them what fiber optics was and what broadband communication was, and then, tried to find out how they might react with respect to their use. It was very interesting.

When they did, most wanted universal service.

Most of them had not thought about telecommunications policy. Most also wanted some form of government regulation and controls that would assure adequate service. Of course, everybody wanted reasonable costs and the vast majority of them mentioned privacy. They were getting very much worried about that.

They were aware, also, of the competition between telephone companies and cable. They thought it would be a good idea to increase that competition, to let telephone companies into the game, but of course, this study was done at the time when the price gouging and the poor service from cable were at their apex, so they think a little differently now.

But what did they think they would get that was most important to them from a universal information network? First of all, they wanted interactive video in emergencies. That was the thing they wanted most. Then they wanted information, educational opportunities and health services.

They were less interested in entertainment, because they felt they had enough. Neither were they interested in home shopping or in financial services. About the lowest thing on the list was movies by demand. Now, it is significant that opinion leaders like this opted for self-protection. I think that is something that we should take into account. They opted for telephone companies, if they were going to be let into the house on a broader basis, that they should continue to be common carriers, even if they were allowed to provide programming. They felt that cable should be forbidden to exclude services or programs that they do not originate or own.

That is just one study, but it does show the way people are thinking, that they are learning something about telecommunication policies. I think the policies that we develop should be a three-fold combination, and I'm talking always for the public interest.

The technology providers, the program providers, the broadcasters, and the cable systems, they can take care of themselves. But I think it is incumbent upon the whole country to re-establish the public interest element of the Communications Act to care for consumers.

The first thing that we need to do is to provide the broadest technical bandwidth possible to the home, and probably, basically through fiber. Second, we need to apply the First Amendment print policy to all content and couple that with common carrier control of the signal originator, whoever that originator may be. Third, we should have guaranteed, open access to the network for all comers. Only at that point, where we have that kind of access, do we let market forces prevail. No matter what they may tell you, market forces do not prevail now, because there is too much ability to keep certain factors and certain elements out, as NBC was kept out of putting on a 24-hour news service because of Ted Turner's leverage with cable owners.

There are two added policy needs that are essential in our society. One is adequate funding of a public communications service that is broader than the Public Broadcasting Service of today, that will use other means of communication than broadcasting. This will go into cable, will go into perhaps records and cassettes and other things that are coming along, and that will provide new educational services. A second social need that we have is to protect the needs and interests of children in this whole communications system. I do not mean phoney protection, the kinds of rules that first were adopted by the Federal Communications Commission as a result of the Children's Television Act, which did not protect children at all. Those of you who follow this, I think, know that stations were coming in and saying that they were doing the right kind of programming for children under the Act, because they were continuing to do all the cartoons that they have done before.

Now, what about needed action? I think that we should have accelerated deployment of a fiber optic network. We should not be held back by telephone company demands that they have the right to do programming, so that after laying fiber, they are assured of a quick profit. And as others have said, I think that there should be other players with these new services, so that there are a good number of possibilities for anybody to get a diversity of sources. The telephone companies should not be allowed to block competition in the local exchange areas where they have a monopoly. They should not be allowed to buy and operate cable systems.

We will have a much healthier system, especially for protection of First Amendment rights, if we have these many highways former Chairman Ferris spoke of—DBS, radio, MMDS, cable—whatever. A thousand flowers is what we need. But I do think, from all that I can see, fiber is going to be the method of choice. It certainly is for trans-Atlantic cables. It certainly is for trunk lines already. We ought to make sure that we get fiber to the curb. The drop may be anything, just so it does not degrade the signal. It should not be so narrow in bandwidth that the everyday consumer can't get the highest quality service that he or she is able to afford.

It worries me that telephone companies are proposing ISDN as a stopgap on the way to high definition television. Subscribers will pay once for a low quality signal, than pay again to upgrade to HDTV. I think we have to look at things like this to make sure that we do not have much less quality than we could have.

We may have less quality for HDTV than we could have because of the FCC policy to protect over-the-air broadcasting. On the other hand, that certainly is good policy in a way, if we can have sets that will take both systems. But we should see that we don't let any technology come in and give us a degraded product.

There are certain things that Congress ought to be doing, and that we ought to try and make them do. If the American people are determined enough, they really can move the Congress, even in this field. Congress should require the FCC to develop a realistic estimate of the cost of deploying the communication highway that we are talking about. Not that the estimate is going to be definitive, but so that we get out of this argument where the telephone company says it is going to cost us a thousand billion dollars, and unless we let them buy out Hollywood and give you all the service, you can't have it. Congress, if they use the FCC properly, can make an informed judgment of the needed roles of the federal government and private interests in developing a universal, open information system. It can make a judgment on how to maintain the pluralism of our present communications system, if fiber is the dominant technology.

We do have to be careful that one industry may own the dominant technology and it may try to keep other technologies from moving content. The President has an important role. I hope we have one who understands the importance of telecommunications, and who will appoint people to the Federal Communications Commission who are competent to do the jobs that they need to do.

The President also has the means for developing a national communication policy through the National Telecommunications and Information Administration, which is his agency to use. He can't force policy, but he can be a pretty good advocate and maybe even get Congress to act. The important thing is to have an FCC where the commissioners are willing to use the broad authority that they have to shape our communications system under Section One of the Communications Act.

We need men and women who are willing to focus on basic policy. A determined Commission through hearings and rule-making can develop concrete policies that would serve the best interest both of the industries involved and the general public. Unfortunately, in my experience, except in very few instances, the FCC from top to bottom, is not much interested in the public interest.

I think we need to change that in our new presidential administration. A strong FCC willing to make policy is desperately needed, especially if Congress continues to muddle along on communication policies the way they have in the past. This holds right up through the latest cable act where they did not make carefully thought-out new policy, but responded to the pressures of powerful interests that tear them back and forth. This keeps them from really thinking through the problems to solutions that will benefit us as consumers.

QUESTION: WHAT ABOUT GOVERNMENT PICKING WINNERS?

From the audience: I have a couple of questions. The first, directed to Charles D. Ferris who said government should not pick winners, but rather should promote variety. I would note the case of AM radio where the government promoted variety and didn't pick a winner and we have a severe market failure. We don't have AM stereo. On the other hand, the FCC is currently involved in trying to pick a winner in HDTV and if they instead do what you suggest and promote variety, no one would ever invest in HDTV with multiple standards. Also, we're involved in global competition where we look to Japan as our primary competitor and to what they're doing in terms of industrial policies and state planning. The Japanese did initially deregulate the telephone company, got intricately involved in trying to develop an information grid, providing services, broadband interactive services to each household and each business by the year 2010 or 2015. I'd like to hear your comments on that, and also, to Eli, who suggested a value-added tax from all, I guess, information providers? I'd like to know what would be taxed, whether it would be revenues or the value of the spectrum used or profits or what? What exactly would it be used for?

Charles D. Ferris: Let me just address the winners and losers issue. I think when it comes to standards setting, you can make a case for the government stepping in at the right time, although government can step in prematurely and pick a standard which really is technologically inferior. But that's a matter of timing. I was talking much more in terms of service providers. Who's going to win or lose? I just don't think that the government should say that the telephone company is going to be the primary provider of telecommunications services, and then, the other alternate providers are going to lose.

If they are going to permit people to lose, it would be that they lose in the marketplace by the fact that the rules were played fairly. There was no use of monopoly power to drive someone out of business with predatory pricing. But if your product didn't compete and you failed for that reason, I think government should tolerate losers in that situation.

Audience question: If government should not be in the business of picking winners, should it ever choose itself to become a winner? For example, should the government upgrade FTS 2000 or 3000 to become a service provider of last resort for, say, the information poor who cannot otherwise afford these new services?

Eli Noam: I don't think government is the actual information provider, as opposed to assurer that information services will be offered, which is a good idea. I don't think there is any particular experience; even FTS 2000, provided the technology. The government contracts for technical facilities offered by private carriers. I would not want to see government becoming some kind of, like the Library of Congress, literally, a kind of information provider. I think it could be an information library that people could access, but not as, in this instance that you've described.

To get to your other question on the value-added tax—this is not something that will happen in the next 10 years or so. But eventually the logic of competition will make it impossible to overcharge some customers in order to undercharge others; it is that kind of logic. We are able to muddle through because there is still a significant amount of monopoly, regulated monopoly, and you can play that through, but it will not last. We are trying to do this through the access charges, but the access charges that are above cost are still predicated on some kind of philosophy like: there's the network, and if you access into the network, you pay. But if you have a network-of-networks arrangement, you don't have the one network, and therefore, access charges will not work. Also, they distort the industry structure because if you're vertically integrated, you don't access and therefore you don't have to pay charges.

It would be some kind of a value-added tax system which will probably be on carriers, mostly. It could be like the value-added tax system that exists in Europe in which you just pay the tax on the incremental value that you add to that particular service or product. This, it seems to me, is a likely kind of future scenario.

But as I said, this is not something that is likely to emerge in this particular political environment for a while.

QUESTION: ARE CONCERNS ABOUT PUBLIC BROADCASTING RELEVANT HERE?

Audience question: I've heard each of you gentlemen today, or on earlier occasions, refer to public broadcasting as an area where there should be additional investment. Currently, there's an annual appropriation of over $300 million in community service grants for public television stations around the country. Virtually none of this goes into educational, instructional, or what we would agree in this room is public service programming. If you were to go, as I think a few of you did, to the public television conferences this year, last year and the year before, the only programming discussion centered around questions like: How do we get something else like "Mystery" and "Masterpiece Theatre" that will pledge well, that will get a big audience? My question is, why do you feel that additional public investment in this current system, where there is very little discussion of public interest or instructional programming, be in the public good?

Everett Parker: Don't get me wrong, I was not calling for more investment in the mess that we have, that we call public broadcasting. The use of the money is a scandal. We all know the difficulties too, of the way that it's run. We need to rethink public broadcasting. I was part of a group that John Wicklein put together, that made pretty good recommendations. But at that particular point, Representative Dingle was sick and tired of public broadcasting and washed his hands of doing anything about reforming the system.

We do need to reform it, but we need public broadcasting as an alternate system, and we need it also to bring things to people, not just to the upper middle class that can afford to make contributions to the stations, but programming to the people who really need services from public broadcasting.

Much more needs to be done for children. It does not look as if we're going to force commercial broadcasting or cable to do any better. We need to have an alternative system.

QUESTION: WHAT ABOUT OTHER NATIONS' TELECOM INFRASTRUCTURES?

Audience question: Charles D. Ferris referred to the superiority of America's telecommunications infrastructure. Is that a historical comment or a projection into the future? For example, the telecommunications and information age report that the NTIA[1] did, suggested that France was way ahead or will be way ahead in the implementation of digital switches and national ISDN. FCC Chairman Al Sikeshas commented about Tokyo, for example, being wired with optical fiber by the year 2000. Such comments imply that others are well ahead of the United States. Are we now talking historically about America's superiority or are we assuming America is currently superior and will continue to be superior over the next five years, including in digital technologies?

Charles D. Ferris: I think, objectively, the U.S. has the most efficient, most effective telecommunications infrastructure in the world. France can have their toys. What they did was to put video screens in every home. But we could do that too, if we wanted to put the cost of that screen in the rate base, as they did. Now, do you want to put $75 on everyone's telephone bill? I don't think it's a very efficient use and allocation of resources to do that. Why did they do it in France? They did it in France because their Yellow Pages™ lost money.

As for digital technology, I think that we have available in our network, the most advanced telecommunications systems, switches and technology and transmission facilities that are needed for the foreseeable future, for what we have to provide in this country. Now, you can gold plate any telecommunications facility. That was somewhat the history when AT&T was the sole provider. I was sort of was fascinated when AT&T came before the FCC wanting to build a microwave tower—it's a little facetious—which could withstand, it seemed to me, a nuclear ground-zero blast.

Audience question: Is fiber to the home another example of gold plating the network?

Charles D. Ferris: I don't know if fiber in the home, which Dr. Parker talks about, can be justified. Can you really justify the capital cost of putting fiber into the home—to every home—when a twisted pair can do the job with the services that the home presently desires? I don't know if you can. I think that adds unnecessary capital costs. It throws things in the rate base which increases, probably only a small amount, but it does raise rates. Is that efficient use of capital expenditures for your telecommunications infrastructure? I don't think it is, at this point. At some point it will be, if we do need another broadband into the home. But now it seems that the Bell operating companies, as the example that Dr. Parker used, are now being able to provide broadband-like services over a twisted pair, with the compression technology that they have—specifically, a video picture over the twisted pair.

If we can have these services with the twisted pair, should we spend your money on more broadband? Sure, you can gold plate anything. I think that we already have the most efficient and effective use of capital resources for the telecommunications infrastructure here in the United States, of any country in the world, bar none.

QUESTION: WHAT ABOUT POLICY MAKING ENTITIES OTHER THAN THE FCC?

Audience question: Most of the discussion in this session has been directed toward what the Federal Communications Commission can or cannot do. With all do respect to a very distinguished and effective chairman of the Commission, who is present, should that be the case? There are limitations on what a regulatory commission can do. We seem to have forgotten that there once was an Office of Telecommunications Policy in the White House. There have been some suggestions of a national commission to look at questions about an *information infrastructure* (the term may imply something broader than the telecommunications infrastructure). Do the issues that have been raised today, and the question of the transformation of our democracy, deserve the attention of the White House? Should there be a national effort to examine these issues of equity in access to information and how we achieve, as Everett Parker has said, the purposes of the First Amendment, which is the diversity of an information marketplace available to all citizens?

Everett Parker: We cannot get anywhere unless the White House takes the lead. You are not going to get an educational program for the country from the FCC. You are certainly not going to get it from the industry. You're not going to get it from the press, which puts these things which are vital to our lives, only on the business pages.

So, to have the White House take the lead is the way to go in trying to develop a sensible, long-range policy. This requires an educational program that will make people understand what they can and can't do when we put in these advance systems, including who would need to use them.

I can't help but sit here and think about the book, *The Wired Nation* (Smith, 1972), whenever I make a plea for equal opportunity for the homes. But if we're going to have broadband communication, eventually, I want to have it for everybody.

Charles D. Ferris: I think there is the likelihood of something happening. Governor Clinton and Senator Gore certainly are much more comfortable with these issues, and address these issues, than any President or administration that I've been familiar with in my 30 years in Washington. They think telecommunications is important and I'm sure that there will be an interest at the top levels of government in these issues. This will be reflected in the people that they put in the key positions, making telecommunications policy during the next administration.[2]

[1]Several parties repeated the oft-stated analogy that telecommunications facilities and services will be as important to the future performance of the U.S. economy as transportation systems have been in the past. See the National Telecommunications Information Administration (NTIA) report on Telecommunications in the Age of Information, Department of Commerce, (1991), p. 21.

[2]The suggestion was made that Stuart Brotman's (1987) paper would be excellent reading on the topic of a new national communications policy.

Chapter 10

EPILOGUE

BY FREDERICK WILLIAMS AND JOHN V. PAVLIK

In broadest terms we have sought to examine the needs for, and applications of, an interactive and eventually multimedia, citizens' information service in the United States. Although as can be seen throughout the pages of this volume, not everybody agrees on the nature of such a service, or even if one is necessary, our aim in this epilog chapter is to draw together the main views of what a National Information Service *might entail. The features of this service have been approached primarily with the aim of best serving the broad range of citizens in our increasingly multiethnic, socially stratified, and evolving post-industrial society. More of our attention has been given to optimizing citizen benefits of such a system and with more attention to broad social and economic benefits than what might be tailored to the interests of large service providers such as publishers or telephone companies. In brief, this project has tried to establish a "high ground," citizen-oriented study of national information services. We have sought an optimum planning model, a vision of "what might be," developed on the assumption that too much of the current activity in this area (regulatory change, business trials, public service demonstrations) has suffered from a short-term, low-risk, emphasis-on-expediency view. Unfortunately, too, the strident arguments among newspaper, telephone, cable and broadcast interests on who should provide information services has overshadowed concern for the needs and interests of everyday citizens. In the largest view, we have considered the vision of how a national information service could transform the nation's communications infrastructure on an order of magnitude that the telegraph, the telephone, and broadcasting have done in earlier times.*

We organize our discussion in response to seven questions:

1. What is a national information service?
2. What is the technology Basis?
3. Are there genuine wants and weeds?
4. What are likely benefits?
5. Is federal policy necessary?
6. Who pays?
7. What's next?

1. WHAT IS A *NATIONAL INFORMATION SERVICE*?

The fundamental premise is that for an "Information Highway" truly to serve our nation, it must be easily available to every citizen who wishes access. It must provide a personal link to those institutions most close to the citizen, namely local schools, health care institutions, and local government. Because it will take a federal initiative to bring these services to every citizen, we refer to this as a "national information service." The proposal proceeds as follows: Consider that we have a new communications medium available to us—one that like the telephone, radio, or television, has become technically feasible, has had early specialized applications, but may eventually explode into wide public use. Further consider that like these earlier media, widespread availability of this new medium could be of great benefit to everyday citizens and, in aggregate, to their society. The new medium to which we refer is the recent combination of computing and telecommunications technologies which can put the multimedia power of interactive voice, text, graphics, image, and moving image communications in the hands of every citizen. This new medium can provide services whereby users can variously request information, perform transactions, send or receive messages, conduct forums, post notices, or "publish" in a general sense of the process. This is the vision of what we are calling a *national information service.*

• Geographically, as well as in terms of policy, *national* in our project has referred to the United States, including relevant national telecommunications regulatory authorities. Although our experience in this project has led to underscoring the importance of the availability of local information resources, we feel it will take national policy to design such services on a widely distributed, available, and standardized design to this country. It will take a national initiative to encourage the service we envisage, although many of the individuals services will have a local character.

• It is also important to anticipate the link of a national information service into international networks much as the U.S.-based Internet has provided an international network for research and other communications.

• *Service* refers to an emphasis upon meeting citizen and institutional needs on a wide-ranging level. We have also assumed a public quality in that the system is accessible to all potential users as is the existing telephone switched network. Accessibility reflects "universal service" and First Amendment policy issues in this country. The emphasis is on collective public needs, not on mainly commercial opportunities for the information provider, although the latter are an important component of the system.

• Although electronic information services in their typical current form may involve access to recorded voice information ("audiotex") or screens of textual information or simple graphics ("videotex" or "teletex"), in more advanced versions they will have "multimedia" qualities. That is, the service of the future will accommodate mixes of voice, sound, text, graphics, images, and moving video images. We should plan for eventual multimedia options. Some discussion of prior services was introduced in chapter 3.

• Again, in the focus of the present project, an electronic information service was also considered as a potential *new public medium* for communication (i.e., "new" in the sense that there would be a national initiative to develop services according to standards that would be widely and inexpensively available to the American public, not unlike national policy initiatives that in the past promoted telephony, radio, and television as widely available public telecommunications services serving in the public interest, convenience and necessity).

• A national information service (NIS) could be both an interactive communication service, in the sense of citizen-to-citizen, citizen-to-institution, and institution-to-institution linkages, as well as an information resource in the sense of providing easy access to public records, files, notices, or messages of benefit to the nation's citizens and institutions. This includes information vital to the democratic process, including information on voter registration, voting locations, candidate information such as speeches, policy positions and voting records, or even electronic town hall meetings. An NIS could also provide as a "universal service" basic connectivity to fire, police, and medical emergency services. The same links could also be the basis for routine utility meter-reading of electricity, water, and gas, and integration of such information into conservation studies.

• For application in public service, a national information service could be designed to serve the nation's public institutions where the value of network services is already becoming well documented in areas of education, health, security, governance, environment, and economic development. Public libraries could be local nodes on the network. It is not difficult to calculate likely economies offered by the network where savings can be transferred into payment for network services. On a similar level, the service could be made available at reasonable rates to commercial institutions who either cannot afford or who choose not to build private networks. If the foregoing can be realized and can be self-supporting, it is likely that a wide range of citizen services can be offered at affordable costs, and with sharing these costs in some form between providers and users. It will also be essential to protect the privacy of individual citizens using a national information service.

2. WHAT IS THE TECHNOLOGY BASIS?

A national information service can be developed by enhancing the platform already provided by the nation's public telephone and possibly cable television networks. Technically, this is a mix of wideband transmission and switching technologies increasingly augmented with access to computer-managed information files, as well as software supporting transaction, message, and publishing systems. The network component of a national information service is not a single conduit but a virtual network that draws from the nation's multiple networks as its physical basis—a "network of networks" that could include components from telephone companies, cable television systems, and specialized network technologies. This marks a radical departure from the past when it was the principle of the scarcity of available communication channels that drove public policy. In the vast information superhighways of now emerging, it is the abundance of channels that should drive public policy. In a broadband environment, there is plenty of room for a national information service available to every home. A critical further technological component are devices supporting user access to the system. Without inexpensive and easily used access devices, we will not have a citizen oriented national information service. Whether the devices be modifications of the telephone, computer terminals, a new type of visual display panel, or a mix of all alternatives, they must be widely available and uncomplicated to users.

• As argued by academic researchers Herbert Dordick and Dale E. Lehman in chapter 8, a national initiative should not suffer from "trickle-down" assumptions; that is, getting access technology in the hands of users may well be a critical starting point. We should not overlook, either, new alternatives for user access devices such as the "digital appliance" described in chapter 2 by Knight-Ridder executive, Roger Fidler.

• It is likely that electronic technologies of modern newspaper publishing and future delivery systems will be a part of the technological underpinnings or a complement to a national information service as outlined by Roger Fidler in chapter 2. We should consider, too, distinctions between "news" and "information" as well as constrasts in social responsibility between newspaper and electronic information services, as discussed by *Seattle Times* publisher, Frank A. Blethen in chapter 6.

• Specific to the service is the need for development of software that could be used to manage the network configuration, manage the various services, and offer the interface with users. User interface technology is especially important for a citizen-oriented national service in that access and use should not require special skills nor expensive purchases (chapters 2 and 8). Recent announcements from a variety of companies suggest that standardized, easy-to-use interface technology is now entering the marketplace. The Intel Corporation and Microsoft announced on May 5, 1993 that they had agreed upon a standard way to integrate the telephone and the personal computer. This newly agreed upon system called "Windows Telephony" is analogous to windows for the PC (Fisher, 1993). Similar announcements have come from other companies, including Bell Atlantic Corporation, US West, MCI, and AT&T.

• It is important, too, to consider implications of the plans of the Regional Bell Operating Companies to upgrade the public network as we move toward what many are calling "the information highway." For example, what are the implications for a national information service of the "Advanced Intelligent Network" described by George Heilmeier, President of Bellcore, in the addendum to chapter 4. The AIN will allow third parties and customers to create services, as opposed to in the past when only the telephone companies could do so.

3. ARE THERE GENUINE WANTS AND NEEDS?

Research literature is abundant that documents the correlation between information access and socioeconomic stratification in this country. When socioeconomic stratification is also linked to being classed as minority, elderly, non-English-speaking immigrant or people with disabilities, information access looms even more as a problem. This problem, moreover, is not always just one of availability, but the capability to take advantage of availability, and even the knowledge and attitudes of what information is important to a given situation. Electronic access surely cannot solve all of these problems, but it can (a) improve availability in time and space over other media; (b) by the use of well designed interfaces make that availability more accessible; and (c) by interactive menus and dialogs improve the client's ability to know what information is relevant to a situation.

• Wants and needs for electronic information services are more difficult to predict for the very poor because those who may need the services the most have the least experience available for evaluation to draw from. There is little opportunity to gather data on system use by the most needy. On the other hand, from existing local services, including so-called "information kiosks," we know (chapter 5) that although food and shelter information are a most frequent request from the poor, remaining types of requests may range across literally hundreds of categories.

• There is a growing public mandate, such as discussed by communication researcher Jannette Dates in chapter 6, that new information systems be developed with a major eye toward those who are increasingly disenfranchised in the information society.

• As William Dutton suggests (chapter 5), the foregoing problems can also be examined in terms of problems of "technology access" as a barrier to "information access." This is a further line of reasoning leading to the importance not just of having the information infrastructure relevant to everyday citizens but putting the access technology in their hands to access that infrastructure. It is, again, an example of the "universal service" argument from telephony policy as extended to public information services.

• Evidence also indicates a public desire for message and transaction services perhaps even more so than simply access to information (William Dutton, chapter 4).

4. WHAT ARE LIKELY BENEFITS?

In broadest terms, increasing the information access of the broad range of U.S. citizens is the greatest policy benefit of a National Information System, namely, the presumption that a democracy will best benefit from an informed citizenry. The negative counterpart of this benefit is that access to information services by only the wealthy or information elite will lead to increased stratification of U.S. society. There are likely benefits for the nation's combined industrial and technological policies as well. Development of a national information service will have direct economic benefits in jobs created to develop and maintain the system. There are indirect economic benefits in the form of new businesses now made possible by the presence of a widely accessible information service. This can include new opportunities for business development in economically depressed rural and inner city areas. Finally, there are the externalities created in the form of providing more opportunities for productivity in the delivery of public services.

• Studies should be conducted to inquire into direct and indirect economic benefits of a National Information Service, much as studies have already been undertaken to assess the benefits in upgrades of our current national telephone network, including implications of migration to wideband telecommunications services. Benefits could include a significant growth in telecommuting, and a reduction in the consumption of newsprint and other paper products, all of which would provide important environmental benefits.

• Studies, or in the short term public hearings, should be conducted to determine priorities in the availability of a National Information System for distance education and telemedical applications. Research literature has been growing on the topic of telecommunications applications in these areas, so new research can build upon existing knowledge.

• All major likely applications of the system can also be subjected to cost-benefit or cost-effectiveness analyses so that policymaking can benefit from databased evaluations of returns on the new investments in the service.

• An antithesis to citizen benefits is that as much as every citizen is linked to, and uses the system, there is the danger of the loss of privacy. It is important that studies be conducted to ensure that as much effort is given to technological solutions to privacy requirements as it is to service applications.

5. IS FEDERAL POLICY NECESSARY?

For more than 60 years a Federal policy of "universal service at affordable cost" has brought the telephone to over 97% of all U.S. households. A similar Federal initiative is necessary if we are to realize the widest possible benefits of a citizen oriented information system. A purely market oriented model will only give us much the same as we have now—namely, interconnection among privileged users and commercial services that constitute a lucrative market. It may also lead to greater consolidation in the $179 billion information business, with large companies getting even bigger, and middle-sized companies disappearing because they are unable to compete (MacDonald, 1993). Ironically, the nature of the new technologies, such as AIN, have the potential to create a much more even playing field where companies of any size could develop new information services. If we leave policy in this area to state utility commissions, we will condemn development of information services to time-consuming, special interest state-level legislative debates and fractionating of purpose. Only a national policy initiated by the Executive branch, Congress, or both, with policy implemented through the Federal Communications Commission, will we have sufficient influence to drive the development of a citizen oriented service.

- As a national and a government instituted service, however, we need not assume that the Federal government would build nor operate a national information service. Federal policy can encourage development of a national service by commercial and shareholder interests in a manner patterned after other applications of national telecommunications policies.

- Alternative models include government-commercial partnerships as was the case for the Communications Satellite Corporation. The role of the Federal government is needed for debate and selection of an overall policy for the service, for encouraging the development and section of standards, and for safeguarding public services applications of the network.

• Although not easy, it is possible to envisage development of a national information service as a joint national telecommunications and industrial development policy, linking new communications and information benefits to jobs and business creation, rural and inner-city development strategies, and national initiatives in education and health care services.

• However, it is important to note that the role of federal policy, including in Public Broadcasting, has failed to benefit disenfranchised populations in this country, as argued by researcher and clergyman Everett C. Parker in chapter 9.

• Again, as we are reminded by for FCC Chairman Charles Ferris in chapter 9, massive government involvement in a citizen information system could threaten individual rights to privacy; policy must take this threat into consideration.

6. WHO PAYS?

In the long term the users pay. They pay in a form of fees for certain commercial services that can subsidize public services; public service agencies pay fees to accommodate their communications needs; and income can be generated from advertising services. In the short term, it will take an initial government investment to plan the service and to fund long-term debt for contractors to enhance existing public telecommunications facilities.

• A high priority for study are alternatives for financing the development and maintenance of a National Information System. Some of these finance policy alternatives are well addressed by telecommunications scholar Eli Noam in chapter 9.

• As mentioned briefly by Roger Fidler in chapter 2, there are many thoughts on how advertising may be located in unobtrusive yet effective ways in interactive visual media formats. We need studies both of alternatives for such development as well as studies of how advertising may fit structurally into the financial model of a national information system.

7. WHAT'S NEXT?

There are many possible scenarios; one could be: The President proposes and Congress passes legislation that calls for a 2-year study of the likely plan for a national information service, possibly as a part of existing commitments for a national technology development policy. The executive office has already demonstrated its interest in this area (see Clinton & Gore, 1993). The National Telecommunications and Information Administration (Executive branch of the Federal government) and the Office of Technology Assessment (U.S. Congress) propose studies of the likely nature of a National Information Service, including its relevance to a national industrial technology policy. Studies should be subcontracted on: (a) applications in public service areas of job-training, health, education, governance, and conservation; (b) small business applications; (c) how to link existing and developing local public information systems with national network standards and resources; (d) creating content for a Natioal Information Service, and (e) methods of financing the development of the system. Major communications services and equipment providers should be invited to participate in studies at all levels and to provide research agendas of their own. Importantly, invitations should also be extended to traditional print, broadcast and cable media to play a lead role in the effort to create content for the NIS. Schools of journalism, communication and telecommunication should be invited to study the nature and impact of an NIS on the democratic process. At the end of the 2-year study, legislation is developed and debated regarding whether, or how, to proceed with building the system.

• Among the specific areas for studies include those (a) to develop applications in public service areas of job-training, health, education, governance, and conservation; (b) small business applications; and (c) on linking existing and developing local public information systems with national network standards and resources.

• Various pros and cons should be considered relative to the role of the Federal Communications Commission in development of new national services. Should the FCC take on the role of encouraging and facilitating innovation, entrepreneurial endeavor, and new job and investment creation, or should there be a bias towards promoting diversity in both conduit and content? See comments by former chief commissioners Alfred Sikes in chapter 7 and Charles Ferris in chapter 9. Development of a universally accessible national information service may present an opportunity to pursue both policy courses, by creating an environment for fostering new communication services by a wide spectrum of information providers and users.

REFERENCES AND BIBLIOGRAPHY

[Editors' note: There are a few more entries than are cited; they are included for their overall importance to the topic.]

Abramson, J. B., Arterton, F. C., & Orren, G. R. (1988). *The electronic commonwealth.* New York: Basic Books.

Alstyne, W. W. (1984). *Interpretations of the First Amendment.* Durham: Duke University Press.

Altman, L. K. (1991, September 25). Plans for electronic medical journal. *The New York Times*, Section A, p. 21.

Andrews, E. L. (1991a, October 8). Court lets 'baby bells' branch out: Companies can sell information services starting right now. *The New York Times*, Section D, pp. 1, 5.

Andrews, E. L. (1991b, October 25). Phone companies could transmit tv under F.C.C. plan: Blow to cable industry, viewers expected to benefit from many more choices - new lines needed. *The New York Times*, pp. A1, D16.

Apportioning the spectrum: Are new approaches necessary? (1991, September). *Intermedia*, pp. 33-39.

Arlen, G. H. (1991). SeniorNet services: Toward a new electronic environment for seniors. *Forum Report* (15) Washington, DC: The Aspen Institute.

Arlen, G. H., (1993a). Consumer online services growth slows to 11% annual rate; Community/BBS formats boom as many systems reconfigure. *Information & Interactive Services Report,* 14(6), 1.

Arlen, G. H., (1993b). "UNET" recruits magazine publishers for new online service; TV guide service debuts next month, others due by summer. *Information & Interactive Services Report,* 14(7), 3.

Associated Press DataStream. (1991, November 4). Industry News Release.

Associated Press DataStream. (1991, November 11). Industry News Release.

Arterton, F. C. (1987). *Teledemocracy.* Beverly Hills: Sage.

Aumente, J. (1989). Online for social benefit. *Forum Report* (12). Washington, DC: The Aspen Institute.

Barber, B. (1984). *Strong democracy: Participatory politics for a new age.* Los Angeles: University of California Press.

Barnouw, E. (1990). *Tube of plenty: The evolution of american television*. New York: Oxford University Press.

Bates, B. J. (1990). Information systems and society: Potential impacts of alternative structures. *Telecommunications Policy,* 151-158.

Benjamin, G. (Ed.). (1982). *The communication revolution in politics.* New York: American Academy of Political Science.

Bertiger, B. (1991, August-September). Iridium - a global personal communications system. *Intermedia,* 41-44.

Bevan, N. (1991, August-September). Who need symbols and icons? *Intermedia,* 45-49.

Billings, H. (1991, October 15). The bionic library. *Library Journal,* 38-42.

Brand, S. (1987). *The media lab: Inventing the future at MIT*. New York: Viking.

Branscomb, A. (1991, September 1). Extending press freedom: The newest "publishers" need no presses to "print" their message. *The Quill* , 20-21.

Brotman, S. N. (Ed.). (1987). *The telecommunications deregulation sourcebook.* Boston: Artech House.

Bruce, R. R., Cunard, J., & Director, M. D. (1986). *From telecommunications to electronic services: A global spectrum of definitions, boundary lines, and structures*. Boston: Butterworth Group.

Carter, T. B., Franklin, M. A., & Wright, J. B. (1988). *The First Amendment and the Fourth Estate: The law of the mass media.* Westbury: The Foundation Press.

Cave, M., Lever, K., Mills, R., & Trotter, S. (1990). Cost allocation and regulatory pricing in telecommunications: A UK study. *Telcom Policy* 6(14).

Clinton, W. J., & Gore A. (1993, February). *Technology for America's economic growth, a new direction to build economic strength*. Washington, DC: Office of the President of the United States.

Cole, B. G. (Ed.). (1991). *After the break-up: Assessing the new post-AT&T divestiture era*. New York: Columbia University Press.

Communications, computers and networks. (1991). *Scientific American* (special issue).

Computer confusion: A jumble of competing, conflicting standards is chilling the market. (1991, June 10). *Business Week*, 72-77.

Coy, P. (1991, September 16). How Do You Build an Information Highway? *Business Week*, 108-112.

Coy, P. (1991, October 7). Super phones: Video hookups, satellite links, laser switches - high-tech miracles are in the making. *Business Week*, 138-144.

Danziger, J. N., Dutton, W. H., Kling, R. & Kraemer, K. (1982). *Computers and politics.* New York: Columbia University Press.

Data chasers: Our panel of experts ponders newsinc.'s proposal for the paper of the future. (1991, March). *Newsinc.*, 47-54.

Davidge, C. (1987). America's talk-back television experiment: QUBE. In W. Dutton, J. Blumler, & K. Kraemer (Eds.), *Wired cities: Shaping the future of communications* (pp. 75-101). Boston: G.K. Hall.

Dennis, E. E. (1988, October). Communications Policy for the '90s. *Communique*, p. 2.

de Sola Pool, I. (1983). *Technologies of freedom*. Cambridge: The Belknap Press of Harvard University.

de Sola Pool, I. (1990). *Technologies without boundaries: On telecommunications in a global age* (E. M. Noam, Ed.). Cambridge: Harvard University Press.

de Sola Pool, I., & Alexander, H.E. (1973). Politics in a wired nation. In I. de Sola Pool (Ed.), *Talking back: Citizen feedback and cable technology*. Cambridge: MIT Press.

de Sola Pool, I., Inose, H., Takasaki, N., & Hurwitz, R. (1984). *Communications flows; A census in the United States and Japan*. Tokyo: University of Tokyo Press.

Designing a universal handset. (1991, August-September). *Intermedia*, pp.53-56.

Donohue, G.A., Olien, C.N., & Tichenor, P.J. (1973). Mass media functions, knowledge and social control. *Journalism Quarterly, 50*, 652-659.

Dordick, H. S., & Fife, M. D. (1991, April). Universal service in post-divestiture USA. *Telecommunications Policy*, pp. 121-128.

Dordick, H.S., & LaRose, R. (1992). *The telephone in daily life; a study of personal telephone use*. Philadelphia: Temple University.

Dupagne, M. (1990, December.). French and US videotex: Prospects for the electronic directory service. *Telecommunications Policy*, pp. 489-504.

Dutton, W. H. (1992a). *Electronic service delivery and the inner city: Community workshop.* Summary: Overview of a Community Workshop held for the Office of Technology Assessment, U.S. Congress, at Annenberg School for Communication, University of Southern California, Los Angeles.

Dutton, W. H. (1992b). The social impact of emerging telephone services. *Telecommunications Policy 16*(5), 377-387.

Dutton, W. H. (1992c). Political Science Research on Teledemocracy. *Social Science Computer Review 10*(4), 502-522.

Dutton, W. H., Blumler, J. G., & Kraemer, K. L. (Eds.). (1987). *Wired cities.* Boston: G.K. Hall.

Dutton, W. H., & Guthrie, K. (1991). An ecology of games: The political construction of Santa Monica's public electronic network. *Informatization and the Public Sector* , *1*(4), 1-24.

Dutton, W. H., Guthrie, K., O'Connell, J., & Wyer, J. (1991). *State and local government innovations in electronic services.* Unpublished report for the Office of Technology Assessment, U.S. Congress: Annenberg School for Communication, University of Southern California, Los Angeles.

Dutton, W. H., Rogers, E. M., & Jun, Suk-Ho. (1987). Diffusion and social impacts of personal computers. *Communication Research 14*(2), 219-250.

Dutton, W. H., Wyer, J., & O'Connell, J. (1993). The governmental impacts of information technology: A case study of Santa Monica's public electronic network. In R. Banker, B. A. Mahmood, & R. Kauffman (Eds.), *Perspectives on the strategic and economic value of information technology investment*(pp. 265-296). Harrisburg, PA: Idea Publishing Group.

Einhorn, T. A. (1981, October 8). Legal issues of in-home banking are many. *American Banker*, p. 14.

Emord, J. W. (1991). *Freedom, technology, and the First Amendment*. San Francisco: Pacific Research Institute for Public Policy.

Entman, R. M. (1992). *Competition at the local loop: Policies and implications*. Washington, DC: The Aspen Institute.

Fiber optics at home: Wrong number? (1991, November 17). *New York Times*, Section 4, p. 18.

Firestone, C. M. (in press). *Television for the 21st century: The next wave.* Washington, DC: The Aspen Institute Communications and Society Program.

Firestone, C. M., & Clark, C. H. (1991). *Telecommunications as a tool for educational reform: Implementing the NCTM mathematics Standards*. Washington, DC: The Aspen Institute Communications and Society Program.

Fisher, F. D. (1992). What the coming telecommunications infrastructure could mean to our family. *A National Information Network, Annual Review of the Institute for Information Studies*. Washington, DC: The Aspen Institute.

Fisher, L. M. (1993, May 5). Standard set for uniting phones and computers. *New York Times*, p: d7.

Forester, T. (1989). *Computers in human context.* Cambridge: MIT Press.

Fountain, J. E., Kaboolian, L., & Kelman, S. (1992, October). *Service to the citizen: The use of 800 numbers in government.* Paper prepared for presentation to the Association for Public Policy and Management, Denver, CO.

Garin, M. N., & Redmond, T. A. (in press). Changing economic structures and relationships among entertainment industry participants in the 21st century. In C.M. Firestone (Ed.),*Television for the 21st century: The next wave.* Washington, DC: The Aspen Institute Communications and Society Program.

Geller, H. (1991). *Fiber optics: An opportunity for a new policy?* Washington, DC: The Annenberg Washington Program in Communications Policy Studies of Northwestern University.

Gherman, P. M. (1991, August 14). Setting budgets for libraries in electronic era. *The Chronicle of Higher Education* , Section A, p. 36.

Gilder, G. (1991, October 14). Now or never. *Forbes*, pp. 188-98.

Glaberson, W. (1993, April 26). Newspapers redefining themselves. *The New York Times*, Section D, pp. 1, 10.

Gore, A. (1991, September). Information for the Global Village. *Scientific American*, pp. 108-111.

Graham, E. (1991, October 29). Plug in, sign on and read milton, an electronic classic: Project Gutenberg is sending good books to computers everywhere - For free. *The Wall Street Journal*, Section A, pp. 1, 14.

Grossman, L. K. (1991, November/December). Regulate the medium, liberate the message: Original intent in the electronic age. *Columbia Journalism Review*, pp. 72-74.

Guthrie, K. (1991). *The politics of citizen access technology: The development of communication and information utilities in four cities.* Unpublished dissertation, Annenberg School for Communication, University of Southern California.

Guthrie, K., & Dutton, W. H. (1992). The politics of citizen access technology: The development of public information utilities in four cities. *Policy Studies Journal* , *20*, 4.

Guthrie, K., Schmitz, J., Ryu, D., Harris, J., Rogers, E., & Dutton, W. (1990). *Communication technology and democratic participation: PENers in Santa Monica.* Paper presented at Association of Computer Machinery's (ACM) Conference on Computers and the Quality of Life, Washington DC.

Guy, P. (1992, October 28). Newspaper's latest link: Dial 511. *USA Today*, p. 01B.

Hanson, W. (1992). The Kiosk phenomenon. *Government Technology*, 9, 16-17.

Hooper, R. (1985). Lessons from overseas: The British experience. In M. Greenberger (Ed.), *Electronic publishing plus* (pp. 181-200). White Plains: Knowledge Industry.

Information & Interactive Services Report. (1993, March 26), p. 1.

Johnson, L. L., & Reed, D. P. (1990). *Residential broadband services by telephone companies? Technology, economics and public policy*. The RAND Corporation, R-3906-MF/RL.

Kahin, B. (Ed.). (1992). *Building information infrastructures*. New York: McGraw-Hill.

Kahn, F. J. (Ed.). (1968). *Documents of American broadcasting*. New York: Meredith Corporation.

Kapor, M. D. (1991, October 17). Building the open road: The NREN as test-bed for a national public network. *Digital Media*, pp. 8-13.

Katsh, M. E. (1989). *The electronic media and the transformation of law*. New York: Oxford University Press.

Levinson, R. W. (1988). *Information and referral networks: Doorways to human services*. New York: Springer.

Lucky, R. W. (1989). *Silicon dreams*. New York: St. Martin's Press.

MacDonald, R. (1993, May 4). *Forecasting new delivery systems*. Paper presented at Newsroom technology: The next generation. The Freedom Forum Media Studies Center, New York.

Markoff, J. (1991, July 3). For Shakespeare, just log on: Large PC libraries are being developed. *The New York Times*, Section D, pp. 1, 5.

Markoff, J. (1993, February 23). Clinton proposes changes in policy to aid technology. *The New York Times*, Section A, p. 1.

Martyn, J. (1986). Database types, present and future. In B. M. Hall (Ed.), *Information in the 1990's: Human perspectives in a machine age*. London: The Library Association Information Services Group.

McCombs, M. E., & Eyal, C. H. (1980). Spending on Mass Media. *Journal of Communication* 30(1), 153-158.

McCombs, M. E., & Nolan, J. (1992). The Relative Constancy Approach to Consumer Spending for Media. *Journal of Media Economics* 5(2), 43-52.

Miller, A., & Rosado, L. (1991, October 28). Dial 1-900 for doctor: Phone services that offer medical advice are the latest trend in health care. But are they too risky? *Newsweek*, p. 48.

Minow, N. M. (1991). *How vast the wasteland now?* New York: The Freedom Forum Media Studies Center.

Money machine: Trading has gone high-tech - And the street will never be the same. (1991, June 10). *Business Week*, pp. 80-86.

Negroponte, N. P. (1991). Products and services for computer networks. *Scientific American*, pp. 106-113.

Neustadt, R. M. (1982). Electronic publishing and privacy. In *The Birth of electronic publishing: Legal and economic issues in telephone, cable and over-the-air teletext and videotext*. White Plains: Knowledge Industries.

Newberg, P. R. (Ed.). (1989). New directions in telecommunications policy. *Volume 1 regulatory policy telephony and mass media*. Durham: Duke Press Policy Studies.

Noam, E. M. (Ed.). (1983). *Telecommunications regulation today and tomorrow.* New York: Harcourt Brace Jovanovich.

Noll, A. M. (1985). Videotex: Anatomy of a failure. *Information and Technology, 9,* 99-109.

Noll, A. M. (1987). The effects of divestiture on telecommunications research. *Journal of Communications, 37*(1), 73-80.

Noll, R. (1988). *Telecommunications regulation in the 1990's* (Publication No. 140). Stanford: Center for Economic Policy Research.

NTIA telecom 2000: Charting the course for a new century (1988, October). (NTIA Special Publication 88-21). Washington DC: Author.

Pavlik, J. V., & Dennis, E. E. (Ed.). (1993). Demystifying media technology: A Freedom Forum Center reader. Mountain View, CA: Mayfield Publishing Company.

Pearson, L. (1991, October 5-6). *The Chugach conference: Communication issues of the 90's*. Anchorage, AK: The University of Alaska.

Pelton, J. N. (1991, September/October). Wideband telecommunications - 2021. *Transnational Data and Communications Report* , 19-24.

Pepper, R. (1988). *Through the looking glass: Integrated broadband networks, Regulatory policy and institutional change 5.* Washington, DC: FCC Office of Plans and Policy.

Pfaffenberger, B. (1990). *Democratizing information: On line databases and the rise of end-user searching*. Boston: G.K. Hall.

Ramirez, A. (1991, November 17). Fiber Optics at Home: Wrong Number? *The New York Times,* Section E, p. 18.

Reply brief for the Bell Company appellants, United States v. Western Elec. Co., Inc. Nos. 87-5388 (D.C. Cir. Sep. 20, 1989).

Rose, M. H. (1979). *Interstate: Express highway politics, 1941-1956.* Lawrence: The Regents Press of Kansas.

Rosen, J. (1991, September). The erosion of public time: Democracy suffers as public discourse gives way to the shrinking soundbite. *The Quill*, pp. 22-23.

Sackman, H., & Boehm, B. (1972). *Planning community information utilities.* Montvale, NJ: AFIPS Press.

Sackman, H., & Nie, N. (Eds.). (1970). *The information utility and social choice.* Montvale, NJ: AFIPS Press.

Scheck, S. (1991, April). Is it live or is it memory? *Technology Review*, pp.13-14.

Schmandt, J., Williams, F., Wilson, R. H., & Strover, S. (Eds.). (1991). *Telecommunications and rural development: A study of business and public service applications.* New York: Praeger.

Schor, J. B. (1992). *The overworked American; The unexpected decline of leisure.* New York: Basic Books.

Sheekey, A. D. (Ed.). (1991, May). *Education policy and telecommunications technologies.* Washington, DC: U.S. Department of Education, Office of Educational Research and Improvement.

Sikes, A. (1991, October 14). *Video technology News*, p. 3.

Small, K., Winston, C., & Evans, C. (1991). *Road work.* Washington, DC: Brookings Institution.

Smith, R. E. (1993). Confidentiality of Electronic Bulletin Boards. *Privacy Journal, 19*(5), 1.

Smith, R. L. (1972). *The wired nation; Cable TV: The electronic communications highway.* New York: Harper & Row.

Stevenson, R. W. (1991, June 6). "Baby Bells" bill passed by senate: Regional telephone companies seeking to make equipment. *The New York Times*, Section D, pp. 1, 8.

Tale of revenge stirs AIDS furor: Woman claims she's trying to infect men, prompting a surge of concern. (1991, October 1). *The New York Times*, Section A, p. 16.

Techno bloom: Making sense of blossoming technologies. (1991, September). *The Quill*,.

Telecommunications (1991, October 8). *International Herald Tribune* (special section), pp. 9-16.

Telecommunications in the age of information. (1991). Washington, DC: National Telecommunications Information Administration (NTIA), Department of Commerce.

Telecommunications in the *Wall Street Journal* reports. (1991, October 4). *The Wall Street Journal*, Section R, pp. 1-16.

Thimbleby, H. (1991, August-September). Telephones now and .... *Intermedia*, pp. 50-51.

Thornton, J. P. Gerlach, G. G., & Gibson, R. L. (1984). Legal issues in electronic publishing, libel. *Federal Communications Law Journal*, *36*(2), 178.

Towards a reformulation of the communications act. (1993, January). *A Report of the Aspen Communications Counsel's Forum*. Washington, DC: The Aspen Institute.

Trauth, D. M., & Huffman, J. L. (1989). A case of a difference in perspectives: The DC Circuit Court of Appeals and the FCC. *Journal of Broadcasting and Electronic Media*, *33*(3), 247-272.

TV Answer inks deal with Hughes for installation of VSAT earth stations. (1991, September 16). *Video Technology News*, p. 3.

United States v American Tel and Tel Co, 552 F Supp 131 (DDC 1982), aff'd sub nom Maryland v United States, 460 US 1001 (1983).

United States v Western Elec Co. (1987, DDC). 673 F Supp 525.

United States v Western Electric Co. (1988, DDC). 714 F Supp 1.

United States v Western Electric Co. (1989-1). Trade Cas (CCH) 68, 619.

United States v Western Electric Co. (1990, DC Cir). 900 F 2d 283.

U.S. Congress, Office of Technology Assessment. (1990). *Critical connections: Communication for the future*, OTA-CIT-407. Washington DC: U.S. Government Printing Office.

Videophone gets into picture as emerging technology in videoconferencing. (1991, September 16). *Video Technology News*, p. 3.

Westen, T., & Givens, B. (1989). *The California channel: A new public affairs television network for the state.* Los Angeles: Center for Responsive Government.

Williams, F. (1990). The coming intelligent network: New options for the individual and community. In*Annual Review at the Institute for Information Studies*. Washington, DC: The Aspen Institute.

Williams, F. (1991a). Network information services as a new public medium. *Media Studies Journal, 5*(4), 137-151.

Williams, F. (1991b). *The new telecommunications: Infrastructure for the information age*. New York: The Free Press.

Williams, F. (1991c, September). The shape of news to come: The Gulf War was an opportunity for TV news to show off, and to raise questions. *The Quill* , pp. 15-17.

Williams, F. (1992a). The intelligent network: A new beginning for information services on the public network. In F. Y. Phillips (Ed.), *Thinkwork: Working, learning, and managing in a computer-interactive society*. Westport, CT: Praeger.

Williams, F., & Hadden, S. (1993). On the prospects for redefining universal service. In J.R. Schement and B.D. Ruben (Eds.), *Between communication and information, information behavior, Vol. 4*. New Brunswick, NJ: Transaction.

Wilson, D. L. (1991, September 11). Testing time for electronic journals. *Information Technology*, pp. 22-24.

Wurman, R. S. (1989). *Information anxiety*. New York: Doubleday.

GLOSSARY

[This Glossary has been developed and updated over the course of many reports and several books. The last version before the present update was in Williams, F. (1991b) *The New Telecommunications.* New York: The Free Press, and is used by permission of the author and copyright holder.]

AIN Advanced Intelligent Network;a network architecture currently in development at Bellcore laboratories that will allow third parties and telephone company customers to customize their own telephone services.

ANALOG Signal representations that bear some physical relationship to the original quantity; usually electrical voltage, frequency, resistance, or mechanical translation or rotation.

ANTENNA A device used to collect or radiate radio energy.

BANDWIDTH The width of an electrical transmission path or circuit, in terms of the range of frequencies it can pass; a measure of the volume of communications traffic that the channel can carry. A voice channel typically has a bandwidth of 4,000 cycles per second; a TV channel requires about 6.5 MHz.

BASEBAND An information or message signal whose content extends from a frequency near dc to some finite value. For voice, baseband extends from 300 hertz (Hz) to 3,400 Hz. Video baseband is from 50 Hz to 4.2 MHz (NTSC standard).

BAUD Speed of transmission in bits per second (bps) in a binary (two-state) telecommunications transmission. After Emile Baudot, the inventor of the asynchronous telegraph printer.

BINARY A numbering system having only digits, typically 0 and 1.

BIT Binary digit. The smallest part of information with values or states of 0 or 1, or yes or no. In electrical communication system, a bit can be represented by the presence or absence of a pulse.

BOC Telephone jargon for "Bell Operating Company," used to refer to divested companies.

BROADBAND CARRIERS The term to describe high capacity transmission systems used to carry large blocks of telephone channels or one or more video channels. Such broadband systems may be provided by coaxial cables, microwave radio systems, or optical fibers.

BROADBAND COMMUNICATION A communication system with a bandwidth greater than voiceband. Cable is a broadband communication system with a bandwidth usually from 5 MHz to 450 MHz.

BYPASS A telephone industry term meaning service that avoids use of the local exchange company network, such as a customer connecting directly into the long-distance network or buying a direct line between offices instead of using the public network.

BYTE A group of bits processed or operating together; 16-bit and 32-bit bytes are common.

CARRIER Signal with given frequency, amplitude, and phase characteristics that is modulated in order to transmit messages. A colloquial use can refer to a telecommunications company.

CATHODE RAY TUBE Called CRT, this is the display unit or screen of a computer or terminal.

CCITT Consultative Committee for International Telephone and Telegraphs, an arm of the International Telecommunications Union (ITU), which establishes voluntary standards for telephone and telegraph interconnection.

CELLULAR RADIO (TELEPHONE) Radio or telephone system that operates within a grid of low-powered radio sender-receivers. As a user travels to different locations on the grid, different receiver-transmitters automatically support the message traffic. This is the basis for modern cellular telephone systems.

CENTRAL OFFICE The local switch for a telephone exchange, sometimes referred to as a "wire center."

CHANNEL A segment of bandwidth that may be used to establish a communication link. A television channel has a bandwidth of 6 MHz, a voice channel has about 4,000 Hz.

CHIP A single device made up of transistors, diodes, and other components, interconnected by chemical process and forming the basic component of microprocessors.

COAXIAL CABLE A metal cable consisting of a conductor surrounded by another conductor in the form of a tube that can carry broadband signals by guiding high-frequency electromagnetic radiation.

COMMON CARRIER An organization licensed by the Federal Communications Commission (FCC) and/or by various state public utility commissions to supply communications services to all users at established and stated prices.

COMPRESSED VIDEO Television systems that require much less bandwidth than commercial television standards; used in business and educational settings.

COMPRESSION A process of condensing audio or video signals or other digital information into fewer bytes of storage so that it requires less sotrage space and bandwidth for transmission.

COMSAT Communications Satellite Corporation. A private corporation authorized by the Communications Satellite Act of 1962 to represent the United States in international satellite communications and to operate domestic and international satellites.

CONVERGENCE A confluence of distinct media technologies into a single electronic computer-driven environment, especially the blending of telecommunications and computers.

CPE Telephone jargon for "customer premises equipment" which may often be distinguished from telephone company owned equipment.

CROSS SUBSIDY In telecommunications, this means that funds from one part of the business (e.g., long-distance) are used to lower prices in another (local service). A controversy is how to prevent cross subsidy between regulated and unregulated parts of the telephone business.

CRT See cathode ray tube.

DATABASE Information or files stored in a computer for subsequent retrieval and use. Many of the services obtained from information or videotext services involve accessing large databases.

DIGITAL A function that operates in discrete steps as contrasted to a continuous or analog function. Digital computers manipulate numbers encoded into binary (on-off) forms, while analog computers sum continuously varying forms. Digital communication is the transmission of information using discontinuous, discrete electrical or electromagnetic signals that change in frequency, polarity, or amplitude. Analog forms may be encoded for transmission on digital communication systems (see pulse code modulation).

DIGITAL SWITCH A telecommunications switch that operates with digital signals (rather than analog).

DIRECT BROADCAST SATELLITE (DBS) A satellite system designed with sufficient power so that inexpensive earth stations can be used for direct residential or community reception, thus reducing the need for local communications by allowing use of a receiving antenna with a diameter that is less than one meter.

DIVESTITURE The break up of AT&T into separate companies.

DOWNLINK An antenna designed to receive signals from a communications satellite (see uplink).

EARTH STATION A communication station on the surface of the earth used to communicate with a satellite. (Also TVRO, television receive-only earth station.)

ELECTRONIC BULLETIN BOARDS (BBS) On-line version of bulletin board where interested readers access messages left by other parties. Most cover specific topics such as international buyers for a product or advertising rates for FM radio stations. Some systems are capable of interactive, "live" discussions.

ELECTRONIC MAIL ("e-mail") The delivery of correspondence, including graphics, by electronic means, usually by the interconnection of computers, word processors, or facsimile equipment.

ESS Electronic switching system. The Bell system designation for their stored program control switching machines.

FAX Facsimile. A system for the transmission of images. It is a black-and-white reproduction of a document or picture transmitted over a telephone or other transmission system.

FCC Federal Communications Commission. A board of five members (commissioners) appointed by the President and confirmed by the Senate under the provision of the Communications Act of 1934. The FCC has the power to regulate interstate communications.

FIBER OPTICS Glass strands that allow transmission of modulated light waves for communication.

FLAT PANEL Best described as a very thin picture tube, it is the digital successor to the analog computer monitor and television screen and combined with microprocessors, memory and communication links make possible "digital appliances," an entirely new class of electronic media.

FREQUENCY The number of recurrences of a phenomenon during a specified period of time. Electrical frequency is expressed in hertz, equivalent to cycles per second.

FREQUENCY SPECTRUM A term describing a range of frequencies of electromagnetic waves in radio terms; the range of frequencies useful for radio communication, from about 10 Hz to 3,000 GHz.

GATEWAY The ability of one information service to transfer you to another one, as when you go from Dow Jones/News Retrieval to MCI Mail.

GEOSTATIONARY SATELLITE A satellite, with a circular orbit 22,400 miles in space, which lies in the satellite plane of the earth's equator and which turns about the polar axis of the earth in the same direction and with the same period as that of the earth's rotation. Thus, the satellite is stationary when viewed from the earth.

GIGABYTE Billion bytes.

GIGAHERTZ (GHz) Billion cycles per second.

HARD COPY Paper version of information.

HARDWARE The electrical and mechanical equipment used in telecommunications and computer systems (see software; firmware).

HDTV High Definition Television.

HEADEND The electronic control center of the cable television system where weaving signals are amplified, filtered, or converted as necessary. The headend is usually located at or near the antenna site.

HERTZ (Hz) The frequency of an electric or electromagnetic wave in cycles per second, named after Heinrich Hertz who detected such waves in 1883.

HYPERMEDIA Information filed and easily accessed by any relation among characteristics of one presentation with details that could be accessed for any of those characteristics. For example, you might be reading a file on computer-assisted design (CAD); a hypermedia interface would allow you to point ("click") to any example in that file so as to furnish more details. Apple Computer's Hypercard is the best known example of hypermedia. Presumably, this is a human-and-machine interface more akin to normal cognitive processes than using a branching menu or access codes for subfiles.

INFORMATION UTILITY A term increasingly used to refer to services that offer a wide variety of information, communications, and computing services to subscribers; examples are The Source, CompuServe, or Down Jones News/Retrieval.

INTERFACE Device that operates at a common boundary of adjacent components or systems and that enable these components or systems to interchange information.

INTERNET A widely used public message network built up of a "network of networks" thus allowing people on different electronic-mail systems to communicate with one another around the globe.

IXC Interexchange Carrier; telephone companies (e.g., AT&T, MCI, US Sprint) that connect local exchanges and local access and transport areas (LATAs) to one another; a highly competitive part of the business.

ISDN Integrated Services Digital Network; a set of standards for integrating voice, data, and image communication; a service now being promoted by AT&T and some regional telephone companies.

K 1,024 bytes of information, or roughly the same number of symbols, or digits.

KILOHERTZ (KHz) Thousand cycles per second.

LAN See local area network.

LASER Light amplification by simulated emission of radiation. An intense beam that can be modulated for communication.

LATA Local Access and Transport Area; a telephone service region incorporating local exchanges, yet usually smaller than a state; typically are serviced by a given telephone company for local services, and interexchange carriers for some intraLATA and all interLATA service.

LOCAL AREA NETWORK (LAN) A special linkage of computers or other communications devices into their own network for use by an individual or organization. Local area networks are part of the modern trend of office communication systems.

LEC Local Exchange Company; the telephone company that supports local calls (nonlong-distance); typically a regulated monopoly. LECs are certificated for areas called LATAs (Local Access and Transport Areas).

LOOP The link that extends from a telephone central office to a telephone instrument. The coaxial cable in a broadband or CATV system that passes by each building or residence on a street and connects with the trunk cable at a neighborhood node is often called the "subscriber loop" or "local loop."

MEGABYTE Million bytes

MEGAHERTZ (MHz) Million cycles per second.

MEMORY One of the basic components of a central processing unit (CPU). It stores information for future use.

MFJ Short for "modified final judgment," an antitrust agreement that set AT&T divestiture in motion.

MICROCHIP An electronic circuit with multiple solid-state devices engraved through photolithography or microbeam processes on one substrate (see microcomputer; microprocessor).

MICROSECOND One millionth of a second.

MICROWAVE The shortwave lengths from 1 GHz to 30 GHz used for radio, television, and satellite systems.

MILLISECOND One thousandth of a second.

MMDS Multichannel multipoint distribution service.

MODEM Short for modulator-demodulator. The equipment that you use to link your computer to a telephone line.

MODULATION A process of modifying the characteristics of propagating signal, such as a carrier, so that it represents the instantaneous changes of another signal. The carrier wave can change its amplitude (AM), its frequency (FM), its phase, or its duration (pulse code modulation), or combinations of these.

MULTIPLEXING A process of combining two or more signals from separate sources into a single signal for sending on a transmission system from which the original signals may be recovered.

NANOSECOND One billionth of a second.

NARROWBAND COMMUNICATION A communication system capable of carrying only voice or relatively slow speed computer signals.

NETWORK The circuits over which computers or other devices may be connected with one another, such as over the telephone network; or one can speak of computer networking.

NODE A point at which terminals and other computer and telecommunications equipment are connected to the transmission network.

OFF-LINE Equipment not connected to a telecommunications system or an operating computer system.

ONA See Open Network Architecture; standards that allow different telecommunications vendors to interconnect with a network.

ON-LINE Being actively connected to a network or computer system; usually being able interactively to exchange data, commands, and information with a host device.

OPEN NETWORK ARCHITECTURE Standards that allow different telcommunications vendors to interconnect with a network.

OPTICAL FIBER A thin flexible glass fiber the size of a human hair which will transmit light waves capable of carrying large amounts of information.

PACKET SWITCHING A technique of switching digital signals with computers wherein the signal stream is broken into packets and reassembled in the correct sequence at the destination.

PC Any home computer ranging from IBM or IBM compatible to Macintosh and Apple lines to Packard Bell, etc.

PCN See Personal Communications Network.

PDA Portable digital appliance, linking the flat panel, microprocessor, memory and communication into a tablet-sized electronic communication medium.

PERSONAL COMMUNICATIONS NETWORK Currently under development, this is a short-range, low power, digital radio link for voice and data terminals that can be accessed through one's personal user number. It provides for portability in local areas, but can be coded to "travel" with the user to different areas. Currently, these services have the prospect of being sold by organizations other than the local exchange company in U.S. markets, an obvious policy issue.

POOLING ("Revenue Pooling") A telephone industry term meaning setting up special collections of funds for intended cross subsidy, as in averaging rates between high cost rural services and less expensive urban ones.

POTS Jargon for "plain old telephone service."

PUBLIC SWITCHED TELEPHONE NETWORK The more formal name given to the commercial telephone business in the United States; includes all the operating companies.

PUC Public Utility Commission, usually the state entity that sets telephone rates.

PULSE CODE MODULATION (PCM) A technique by which a signal is sampled periodically, each sample quantized, and transmitted as a signal binary code.

REGIONAL HOLDING COMPANY (RHC, RBOC) The companies formed to take over the individual Bell system operating companies at divestiture; there are seven (e.g., Pacific Telesis).

RBOC See Regional Holding Company.

SLOW-SCAN TELEVISION A technique of placing video signals on a narrowband circuit, such as telephone lines, which results in a picture changing every few seconds.

SOFTWARE The written instructions which direct a computer program. Any written material or script for use on a communication system or the program produced from the script. (See hardware, firmware.)

TARIFF The published rate for a service, equipment, or facility established by the communications common carrier.

TELCO Jargon for "telephone company."

TELECOMMUTING The use of computers and telecommunications to enable people to work at home. More broadly, the substitution of telecommunications for transportation.

TELECONFERENCE The simultaneous visual and/or sound interconnection that allows individuals in two or more locations to see and talk to one another in a long-distance conference arrangement.

TELEMARKETING A method of marketing that emphasizes the use of the telephone and other telecommunications systems.

TELETEX The generic name for a set of systems which transmit alphanumeric and simple graphics information over the broadcast (or one-way cable) signal, using spare line capacity in the signal for display on a suitably modified TV receiver.

TERMINAL A point at which a communication can either leave or enter a communication network.

TERMINAL EMULATOR Use of a personal computer to act as a dumb terminal; this requires special software or firmware.

TIMESHARING When a computer can support two or more users. The large computers used by the information utilities can accommodate many users simultaneously who are said to be timesharing on the system.

TRANSPONDER The electronic circuit of a satellite that receives a signal from the transmitting earth station, amplifies it, and transmits it to the earth at a different frequency.

TRUNK A main cable that runs from the head end to a local node, then connects to the drop running to a home in a cable television system; a main circuit connected to local central offices with regional or intercity switches in telephone systems.

TWISTED PAIR The term given to the two wires that connect local telephone circuits to the telephone central office.

UNIVERSAL SERVICE Traditionally defined as making voice telephone service easily available at affordable cost. In the coming years, it could be broadened to include other telecommunications services.

UPLINK The communications link from the transmitting earth station to the satellite.

VIDEOTEX The generic name for a computer system that transmits alphanumeric and simple graphics information over the ordinary telephone line for display on a video monitor.

WATS Wide Area Telephone Service. A service offered by telephone companies in the United States that permits customers to make dial calls to telephones in a specific area for a flat monthly charge, or to receive calls collect at a flat monthly charge.

INDEX